Lesser Worlds

Lesser Worlds

BY NESTA PAIN

ILLUSTRATED BY J. YUNGE BATEMAN

Coward-McCann, Inc.
New York

ACKNOWLEDGEMENT

I FIRST became interested in insects as a result of reading *Social Life in the Insect World* by J. H. Fabre, and I have gone on reading the works of Fabre with very great pleasure ever since. I am sure it is not too much to say that no naturalist has ever carried out such detailed and imaginative experiments as Fabre did, or observed the lives and habits of insects with such devoted care, or recorded them with such charm. I owe him a great debt, as any writer on insects must. Apart from *Social Life in the Insect World* (published by Messrs. Duckworth & Co. Ltd.) already mentioned, I was much helped by *The Life of the Spider*, *The Hunting Wasps*, *More Hunting Wasps*, *Bramble Bees and Others*, and *The Glow Worm and Other Beetles*, all by Fabre and published by Messrs. Hodder & Stoughton Ltd.

I am also greatly indebted to Dr. W. S. Bristowe's fascinating book *The Comity of Spiders* (Messrs. Bernard Quaritch Ltd.), and to Professor W. M. Wheeler's immensely learned and detailed work on *Ants* (Columbia University Press). It is not possible to mention all the other books and papers I consulted, but I should like to record my debt to the charming book on *Wasps Social and Solitary* by the American naturalists Mr. and Mrs. Peckham (Messrs. Constable & Co. Ltd.); *British Ants* and *Guests of British Ants* by H. St. J. K. Donisthorpe (Messrs. Routledge & Kegan Paul Ltd.); *Insect Natural History* (Messrs. William Collins Sons & Co. Ltd.) and *Social Behaviour in Insects* (Messrs. Methuen & Co. Ltd.), both by A. D. Imms; *Ants and Men* by C. P. Haskins (Messrs. George Allen & Unwin Ltd.); *Insects at Home* by J. G. Wood (Messrs. Longmans Green & Co. Ltd.); and *The Social Insects* by W. M. Wheeler (Messrs. Routledge & Kegan Paul Ltd.).

I am very grateful to Dr. Bristowe, who was kind enough to read the sections on Spiders and the Solitary Wasp in manuscript and to give me a great deal of help and advice; also to Dr. O. W. Richards of the Imperial College of Science and Technology, and to Mr. E. B. Britton and Mr. G. E. J. Nixon, both of the British Museum (Natural History), who read other sections of the book in manuscript and were most helpful in the criticisms they were kind enough to give me.

NESTA PAIN

CONTENTS

Ants

SPIDERS

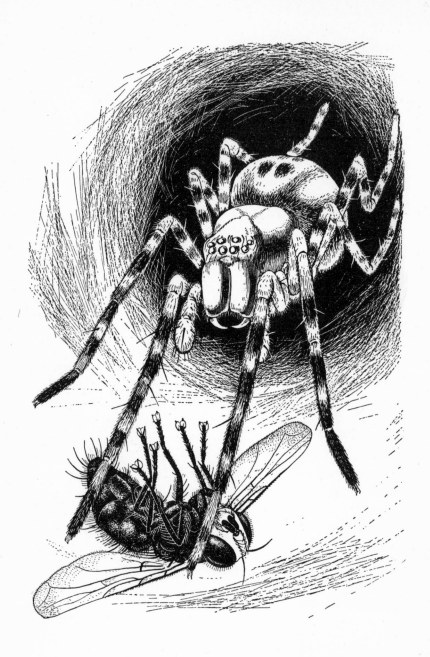

1

THE HUNT

IT is related that Domitian 'delighted to see the jollie combats betwixt a stout Flie and an old Spider.' Unless the spider was disabled in some way, the combats must have been distinctly one-sided and it is unlikely that they were very jolly for the fly. Spiders, in fact, are so able in securing and dispatching their prey that in a single year the insects consumed by the spiders of England and Wales weigh more than all the human inhabitants of these countries put together. This fact might well incline us to such useful creatures, for without them we should certainly be plagued by flies and moths and other small insects to a degree which can hardly be imagined. None the less, there are few people who feel any emotion warmer than a mild indifference for spiders, and there are many who feel a positive loathing and distaste for them.

However, it has often been remarked that there is nobody so ugly, nobody so cold-hearted and unscrupulous that he has not at least one admirer to sing his praises, blind to every defect. The spider has Dr. Muffet.

I will declare unto you the rich virtues and good gifts of Spiders (he writes). The colour of a spider is somewhat pale, such as Ovid ascribeth to lovers, and when she hangeth in her web, with her legs wide and large spread abroad, she perfectly and lively expresseth the shape and proportion of a painted starre. The skin of a spider is so soft, smooth, polished and neat that it farre surpasseth that of the daintiest and most beautiful strumpet's. She hath fingers that the most gallant virgins desire to have theirs like them—long, slender, round, of exact feeling. Indeed, there is no man, nor any creature, that can compare with her.

The author of this somewhat surprising description of the spider was writing towards the close of the sixteenth century and was the father of the Miss Muffet who was so greatly alarmed by the unexpected appearance of a spider while she was eating her curds and whey. Perhaps her father's unusual affection for these creatures often exposed her to such encounters. However that may be, there is no doubt that, although Dr. Muffet's admiration is exaggerated, the spider does possess undeniable virtues as a destroyer of insects.

The insecurities of life for a small insect, inadequately armed, are horrifying beyond the reach of imagination. One single acre of grassland may hold two million spiders, two million jaws hungry for prey. Imagine the journey of an insect through this jungle. Wherever you crawled, if you were an insect, whichever way you turned, there would be one spider lurking to kill in every three square inches of your journey. She might be below the ground, waiting at the door of her hollow tube to leap upon the unwary traveller and drag him down into the darkness of her home, where she could devour him at leisure. She might be hunting on foot, fast and silent, following close at your back; or she might be standing hidden behind this blade of grass or the next, giving no warning, making no sound until the final spring hurled her on top of you. It would be too late for flight by then. Or she might squirt her gum over you from a distance before you had even seen your enemy. You would be lying helpless and bound. You would have to look on without hope of rescue as she made her leisurely approach to inspect her catch and enjoy her feast.

In that world of menace, it is safer to fly than to crawl. On the ground every movement gives away the traveller and escape is difficult. It may be for this reason that hopping and flying insects have evolved, for their chance of survival would certainly be greater in that spider-haunted acre of grassland. Attack, however, generally manages to keep pace with defence, and the spider has her own method of dealing with the insect which approaches through the air. She weaves her gossamer web, made of silk so fine that the silkworm's thread is gross by comparison, and waits for what may come to her.

Webs are not all of the same design. There are the *Agelenid* sheet-webs which can often be seen spread across low-growing plants or smothering the corners of an outhouse. The insect which blunders into this web finds it difficult to crawl on its soft, yielding surface and impossible to jump to freedom. In addition, trip threads impede its progress and the web tilts upwards at the edges.

When an insect is trapped, the spider runs up to inspect what chance has brought her and then decide on the appropriate course of action. A small insect can be dragged back to her lair to be eaten without delay. If the insect is fairly large, she may decide that it is best to weaken it by one or two quick bites and then wrap it in silk before attempting to drag it away. If the insect is really very formidable, however, the spider inspects it with great care and occasionally she comes to a decision which must, one feels, cause her considerable pain. She may judge it prudent to forgo this rather dangerous feast and she may cut the web and let her alarming prisoner go free.

The scaffolding web is rather more elaborate. It consists of a number of threads running in all directions and conforming to no particular design. Some of these threads are smeared with a sticky substance and one or two of them are generally attached to the ground. These are pulled tight so that when an insect comes into contact with one of them it breaks off and the insect is thus lifted into the air. As it struggles, it probably knocks against other sticky threads and is then held more tightly than ever. The waiting spider may run down to her prey, but generally she hauls it up to the main part of her web. When this is done, she turns round and scatters more sticky threads over the captured insect, directing them with her hind legs. The victim, now thoroughly helpless, is bitten for good measure before being dragged to the spider's larder.

The meshed webs of *Ciniflo* and *Dictyna* are irregular structures arranged, as a rule, round crevices in walls or fences or the tops of plants. A comb on the spider's hind leg has teased each thread into adhesive lace which entangles the legs of any unwary insect which happens to step on it. But these meshed webs are unimpressive efforts compared to the much admired orb-web which is built by the Garden Spider, *Araneus diadematus*.

She begins by standing on tip-toe with her body raised so as to allow the silk which she squeezes out from the silk-glands on her belly to be carried away on the air. Sooner or later the strand of silk anchors itself to a branch or a leaf near by. Immediately, she pulls it tight and makes her way across her frail bridge, strengthens it and makes it fast. Now she has the essential starting point from which she can construct her web. She quickly fills in the framework, which may be a simple triangle or something more complicated, lays down the spokes or radii and then fills in the spiral threads. First she spins a few close together to strengthen the centre of the web. Then she puts in a rough, temporary scaffolding to help her before finally setting to work on the evenly spaced spirals which are characteristic of this type of web. These spirals, the last to be woven into the pattern, are smeared with gum which breaks up into small droplets and will hold an insect captive. They do not incommode *Aranea* herself as her feet are slightly oily.

This ambitious structure is only designed to last for a day. Nightly in summer, *Aranea* destroys her handiwork, preserving only the frame, and builds it all over again. This takes her perhaps half an hour, and it is fascinating to watch her at her work. She shows the unhurried dispatch of the skilled operator, turning, moving forward, turning again, with never any need to pause or retrace her steps.

Some orb-weavers provide themselves with a signalling thread which leads to a retreat where the spider lies hidden. There she sits, the line firmly grasped in her claws, and waits to feel the vibrations which tell her that an insect has arrived. Other species spread themselves in the centre of the web to wait. But although methods may differ, the web may be depended upon to provide plenty of game. It is the spider's answer to the evolution of the insect which flies or hops.

The careless moth, the busy fly may well fail to perceive the faint gleam in the sunshine which is all that betrays this carefully-designed snare. If they see it at all, they see it too late, when the nightmare is already upon them. With their wings sticky and their feet sinking into softness, each frantic struggle only serves to make them more securely a prisoner and to signal their arrival to the waiting spider. It is everybody's nightmare come true—the pursuer close behind and you powerless to run away.

Even in this desperate situation, however, all may not be lost, for there are certain kinds of moths, flies, bugs and beetles which are fortunate enough to possess a taste that is highly unpleasant to spiders. Through the slow process of evolution, some of these have found what they must do and this instinct is born again to each new generation. As the spider comes gliding across the web to them, they lie still. They see the monster looming over them with her hairy legs and her swollen belly and still they make no movement. They lie motionless while she peers at them with her eight gleaming eyes that yet see so little, and they do not seek to escape.

The spider bites. It is a savage, poisoned bite, but the wisdom learnt of thousands of years still holds. The insect lies still.

The spider, revolted by the taste, throws the insect to the ground, and then staggers clumsily to the side of the web and is sick. Fluid flows from her mouth and she wipes her jaws against a leaf. The insect is injured, but not mortally, and may survive.

In human terms, the bitten insect has shown the most extraordinary fortitude. It is as though a man should allow his enemy to stab him without resistance or any sign of pain, hoping that the knife will not injure a vital part, gambling on the chance that he will not strike again.

In the case of these insects, we must suppose that instinct operates rather than courage or widsom, and that this instinct has developed through the process of natural selection. For those unpleasant-tasting insects which do *not* behave in this way have small chance of surviving to breed their like. The insect which struggles is often bitten over and over again until it dies, for there is something about the sight of its struggles which provokes the spider to go on attacking it. Even if the taste, when she bites, is absolutely repellent to her, she may keep on biting until the insect at last is still.

Other insects have developed other means of protection from their enemies. Some have found safety by growing a sort of armour-plating. The woodlouse or the beetle, for instance, draws in its legs and waits placidly for the attacking spider to realise that any attempt to make a meal of it will only jar her fangs. Other insects go in for camou-flage, but this is more use in protecting them from birds than from spiders, for the spider is generally much too short-sighted to rely on her eyes in hunt-ing. She can distinguish movement, but subtleties of colouring are lost on her. In any case, even the most skilfully-camouflaged insects must move sooner or later, and if they move into a spider's web she will not be deceived. An examination by touch will soon tell her all she wants to know about her catch.

However, movement itself has been used as a form of camouflage. Spiders are by no means safe from attack by their own kind—indeed there is one sort of spider which lives almost entirely by capturing and devouring smaller or less well-equipped spiders. Some of these unfortunate spiders have hit on the ingenious device of disguising themselves as ants, which possess a flavour so nauseating that they are seldom attacked or eaten by spiders or insectivorous enemies. These spiders are often very much like ants in appearance, and they carry out their impersonation with considerable histrionic powers. Indeed, so vivid is their imitation of the demeanour of an ant, so convincing their reproduction of its characteristic movements, that even expert naturalists are often deceived.

Some spiders go even further. The English naturalist Dr. Bristowe relates that he was walking one day along a path near Rio de Janeiro when he noticed what he took to be a large-headed ant. After watching it for some time, he put a glass-topped box over it. He was then able to examine it and found that it was not an ant at all, but a black spider which was holding the hollow skeleton of an ant over the top of its own body, like a sort of protective umbrella.

Not all spiders build webs—some depend simply on speed and agility to provide themselves with food. The longer-sighted families of hunting spiders concentrate their battery of eight eyes on an insect which has betrayed itself by an incautious movement. They keep it in sight until it has settled and then they creep quietly nearer until a leap will finish the hunt. They need not come very near for some of these spiders can jump as much as eight inches.

The shorter-sighted hunting spiders operate mostly by night. They creep along with consider-able stealth, their front legs stretched inquiringly in front of them, ready to leap into violent action at the first touch of an insect. *Scytodes* is deceptively languid, but she has a special weapon of her own. If she sights prey about a quarter or half an inch away, she gives a convulsive movement and the wretched victim finds itself coated in gum which pins it to the ground.

Half-way between the hunting spiders and the snare-builders come the spiders which hunt by means of silken tubes. *Atypus* builds a closed tube about seven to ten inches long, three-quarters of it below ground. The quarter projecting above the ground serves as the spider's lair. There she waits, her long curved fangs ready for action, until an insect is unwary enough to walk across the top of her tube. At once the unhappy creature is seized and bitten and dragged inside the spider's strong-hold. When it has been eaten at leisure, the spider emerges to repair the hole and await her next victim.

Segestria builds her tube of silk in the crevices of a wall or a rock face. Her arrangements are rather more ambitious. A number of threads radiate outward from the open mouth of the tube and are firmly attached to the ground. These not only keep the entrance stretched wide, but warn the spider of an approaching insect. She sits safely hidden in her tube, with six of her legs holding six of these signalling lines. If an insect knocks against one of them, she can gauge its position at once and rushes out to seize it. She then hauls the insect back into her den, a manœuvre which is not only convenient to her, but has a practical usefulness as well, if the insect is well-armed and able to bite or inflict a sting with its tail. The spider holds it firmly by the middle and retreats backwards into the narrow tube. Inevitably the victim's head and tail are bent away from the spider where they can inflict no damage. Occasionally, however, the catch may turn out to be too formidable altogether and when this happens the spider prudently gives up the struggle and retreats through her open back door.

These signalling threads are rather similar in function to those used by some of the web-builders which have been mentioned earlier. The great French naturalist Fabre decided one day to find out whether these web-building spiders depended entirely on their 'telegraph wire'—the thread which leads up from the centre of the web to their lair—to inform them when a victim had arrived, or whether they used their eyes as well.

He caught a dragon-fly, which he thought would be certain to tempt the spider, and put it in the web.

At once it began to struggle and the whole web vibrated violently. The spider, hidden some way above in a clump of cypress foliage, came running down her telegraph wire at once, discovered the dragon-fly, bound him up in silk and then went home by the same route dragging her prize behind her. 'The final sacrifice will take place in the quiet of the leafy sanctuary,' remarks Fabre.

A few days later, he carried out the same experiment again, but with one small difference. This time he first cut the thread which led to the spider's hiding place. The dragon-fly was large and a 'very restless prisoner', but the web was shaken and the dragon-fly struggled to no effect—the spider did not come down all day. Since her telegraph wire was dangling useless, she received no news of her capture and the prisoner remained uneaten in her web—'not despised', as Fabre explains, 'but unknown'. When darkness fell, the spider descended to make her routine visit to her web which she found torn and in ruins. And in the midst of the ruins she found the dragon-fly. After her hungry day, she ate him on the spot and only when her meal was finished did she turn to the repair of her web.

As Fabre remarks, the signalling thread is something more than a mere bell-rope which communicates a given impulse, but gives no further information. The web is often shaken by the wind, but the spider takes no notice of that. She does not come running down to see if she has made a capture. When an insect arrives in the web, however, she knows it at once. 'Clutching her telephone wire with a toe,' writes Fabre, 'the spider listens with her leg,' and she is evidently capable of decoding the messages which come to her and interpreting them correctly.

2

COURTSHIP

IT may be unfair to say that the female spider has no thought but for food—it cannot indeed be maintained with any plausibility that she thinks at all. Her actions, however, make it clear that the approach of a moving creature small enough to be devoured is more likely to suggest food to her than any other more tender considerations.

Herein lies the male spider's problem. Clearly he must approach the female or matters cannot proceed at all. On the other hand, he is smaller than she is, and the fact that she is well capable of seizing and devouring him makes courtship a matter calling for some courage. He would prefer, naturally enough, not to become 'one flesh' with his mate in this crudely literal sense. And even if he cannot escape this unseemly end, it is in the interests of Nature and the future of the race that the sacrifice should occur after, rather than before, mating is accomplished.

The male spider, then, must find some means of making it clear that he is to be regarded as a mate rather than a meal before he has approached close enough for a misunderstanding to be fatal.

16

If he belongs to a web-building species, he approaches the female by her web. He mounts it with great care, for he knows well that the least vibration will bring the owner down, her mind fixed on prey. He does mean to call her down, it is true, but after his own fashion.

He establishes himself on the web and gently plucks at a thread with his claw. Vibrations go trembling down the line to the female, but his touch is studied and the vibrations are special. They advertise a mate, they declare his intentions. They are, in fact, a message of love—or, at any rate, of courtship. He twangs the silken thread again and again with growing confidence. The female, after a time, responds, and back and forth over the shining strands pass these telegraphic intelligences, these tender morse messages.

Little by little, the male draws nearer, stopping cautiously every now and again to renew his signals. He knows very well that he would be unwise to risk a precipitate approach. At last the female sidles into view. And as that ponderous form looms close beside him, the male spider seems to lose his nerve—he panics and drops down on a thread.

Soon, however, he seems to gather courage again. Perhaps he has some means of knowing that she is inclined to him after all. At all events, he winds up his thread and takes his stand once more upon the web. His aim is to coax her on to a special bridge line that he has woven free from gum and entanglements. She comes to meet him. He draws nearer by slow and cautious stages, and at last **he** stretches out a leg and touches her caressingly. If this advance is well received, he soon grows bolder and proceeds to tickle her with both his forelegs. (Tickling is rather a feature of the courtship of spiders). The result produced by this manœuvre is generally satisfactory to the male spider and, in the case of one species, it is dramatic, for the female is said to pass straight into a trance. The male spider appears to take this as a matter of course and proceeds with his business unmoved. The female recovers consciousness only after the act is accomplished; which seems a little disappointing for her.

One orb-weaving male, *Meta*, hangs about on the outskirts of the web until he sees that the female is occupied in dealing with a captured insect and then seizes his opportunity to approach and caress her. No doubt he reasons—if he reasons at all— that she is too busily occupied with other matters to spare much energy for resistance, and she will be the less likely to want to eat her mate if she has other food conveniently at hand.

For the male hunting spider there is no web by which he may signal his approach to the female, so he has been obliged to fall back on other methods of preparing her mood to receive him. The longer-sighted kinds have evolved means of signalling their identity from a safe distance, but their signals vary from something which may fairly be called a dance of love to a semaphore code.

Lycosa lugubris, for instance, begins to dance

as soon as he comes in sight of the female. He takes up a pose in front of her, his body stiffly erect and his front pair of legs slightly raised and stretched out to the side. Then he slowly raises one of his striking black palps, or arms, in a series of jerks until it is stretched right up above his head. He holds this position—it is undeniably impressive. Then he raises his other arm in the same way. For a moment, he stands poised, both arms raised stiffly above his head, then he brings them down, gently and seductively waving his front pair of legs as he does so.

After a moment's pause he begins again—the whole procedure may be repeated over and over many times for several minutes, until at last he feels this stationary form of wooing has been sufficiently prolonged. Choosing his moment, he makes a sudden rush at his dangerous mate, running sideways round her and rapidly tickling her with his front pair of legs as he does so.

Lycosa saccata, on the other hand, does not take nearly so much trouble with his performance. He merely signals with his arms, raising one and lowering the other, and then reversing their positions. He perseveres with this unambitious exercise, gradually advancing on the female by slow and cautious stages, until she either catches his excitement, or else loses patience and drives him away. His efforts are no more decorative than signalling with a flag, but at least they are unambiguous and adequate to their purpose.

Euophrys frontalis goes in for mesmerism. He is an ornamental creature—his name pays tribute to his 'beautiful eyebrows.' His whole face, in fact, when seen from the front, has considerable claim to beauty for his large bright eyes are fringed with rust-red and set against a black background. By contrast his palps are yellow. He makes good use of this colour contrast by vibrating his palps in front of his face in the course of his dance.

He begins his performance as soon as he comes near a female by raising his handsome front legs with a jerk. (They are unusually large and hairy and jet black all the way down, except for cream tips). He holds them above his head for a moment, and then gradually lowers them almost to the ground . . . then snaps them up again. He repeats these mesmeric movements over and over again, occasionally varying the tempo by making two snaps in quick succession. At each upward snap of the legs, he draws a little nearer to the female; and all the time his yellow palps, turned slightly inwards, are moving gently and intriguingly about in front of his black face.

The effect is quite literally fascinating. The female, even when she remains cold to further advances, seems to find it impossible to turn her eyes away from the dance. It is clear that she is deeply affected by the performance—sometimes, in fact, she is even moved to give a rough imitation of the dance herself.

Tarentula accentuata is perhaps a little emotional

in his approach—it is difficult to say whether it is timidity or excitement which causes the violent trembling which appears to affect his limbs. Since the response of the female of this species is often disturbingly aggressive, it may well be the former.

As soon as he sights a female, *Tarentula accentuata* paws the ground with quivering legs. Then, raising himself up on his second pair of legs, he assumes a curious pose with his front legs and arms raised above his head, but curved forwards like hooks. With a jerk, he forces his legs still higher and then lowers them to the ground. They are trembling really violently by this time. None the less, he may advance a step or two before repeating the performance so long as the female is some distance away. At all events, he always repeats it over and over again, sometimes quite literally for hours on end, all the time circling cautiously round the still menacing female. His caution is understandable for as soon as he comes within striking distance, so to say, she will often make a fierce rush at him. Luckily, although frailer, he is the more agile of the two and always just manages to elude this unfriendly attention. Nervous but game, he takes up his display again, in the hope of taming her angry spirit. In this he at last succeeds. He draws close enough to be able to stroke her gently with the very tip of his front legs. After this, in the words of an expert, 'she either drives him away or else, after a violent interplay of front legs, copulation ensues.'

Ballus depressus, in spite of his name, has rather a gay approach. He has abnormally large front legs which are coal-black, relieved by pale yellow tips. He exhibits them by drawing them in a little and sways in a distinctly drunken roll from side to side, staggering a few steps first this way and then that as he does so. Since the species persists, one supposes that this approach must eventually be successful.

One of these long-sighted hunting spiders seems to feel that even with a preliminary courtship dance he runs considerable risk in mating and prudently waits until he sees the female engaged on a meal before attempting to woo her. *Pisaura listeri* tackles the problem more directly. He is a realist and does not balk at the female's preoccupation with her stomach. Alone among spiders he has hit upon the idea of catching an insect, wrapping it up in a parcel of silk and presenting it to his chosen mate before attempting further intimacies. One might almost suppose that he hopes she will still be occupied with it when mating has finished—that moment of greatest danger—so that he can make good his escape; or that it will, at the least, have taken the edge off her appetite.

If mating is a dangerous affair for the long-sighted hunting spider, it is far, far worse for the short-sighted hunters whose approach is unheralded by signals. They must run alarming risks in order to get to grips with the female. The Crab Spider grips her quite literally. He seizes her front legs in his jaws and then does his best to dodge her spirited counter-attack during the second or so it takes her to realise what he wants. He caresses her until her struggles subside and then ties her to the ground with silken cords before proceeding further—a wise precaution, it would seem.

Drassodes takes possession of an immature female, shuts her up in an enclosed cell and mates with her as soon as she is mature, but before she has developed her full vigour—while she is still weak, in fact, from the exhausting operation of changing her skin. This seems hardly gallant, by human standards, but again it is prudent. *Drassodes* is particularly fierce and the adult female is larger than the male. Indeed, the danger of being eaten alive is by no means negligible and the male spiders of these short-sighted species have no means of turning the female to gentler mood while still keeping at a safe distance.

This tendency of the female spider to make a meal of her mate as soon as his immediate usefulness in other directions is over seems peculiarly repulsive to all but the professional observer, who has his own way of looking at these matters. One modern naturalist, at any rate, takes a distinctly robust view of the procedure.

The practice is far from universal (he writes), and in any case it takes place chiefly towards the close of the mating season so that the future of the species is not jeopardised. Though it may seem cruel, it is an economical practice. The male's function is performed and thenceforth he must be regarded as a competitive 'bread-winner.' Furthermore it is established that most males will die a natural death soon after the mating season is passed, so unless he is eaten, his body will be wasted.

The male spiders, in fact, must provide a meal with their bodies when the appropriate season for their other services is past and gone. It is a point of view which is undeniably logical.

However, by no means all female spiders are hostile to the male and some will even live at peace, if not at close quarters, with their mates. This seems no great achievement, but Dr. Muffet, always a vigorous, if prejudiced, partisan of the spider, managed to find cause for ecstatic admiration even in their family life.

I must not pass by their temperance that was once proper to Man (he writes), but now the Spiders have almost won it from them. Who is there now (if age will let him) who will be content with the love of one? And doth not deliver himself up, body and soul, to wandering lust? But the spiders, so soon as they grow up, choose their mates and never part till death. Moreover as they are most impatient of co-rivals, so they set upon any Adulterers that dare venture upon their Cottages, and bite them and drive them away, and oft-times justly destroy them. Nor doth any of them attempt to offer violence to the female of another, or to assault her chastity. So great command have they of their affections, so faithful and entire are they in their conjugal love, like Turtles.

It is hard to understand how this pious myth can have arisen. Certainly Dr. Muffet cannot have observed them with very much care, for even when the male escapes the more savage attentions of his mate and the sexes live together without incident, all is not as idyllic as it might seem.

Dr. Bristowe has observed that in late summer every female of the species *Meta segmentata* seems to be attended by a male spider which lurks cautiously on the outskirts of the web. He was impressed by this apparent faithfulness, but decided to find out whether it was as genuine as it seemed. Accordingly he put a mark on the male spider in ten different webs. Four days later he inspected the webs again and found all the females attended by one male spider as before. An inspection of the marks he had put on the males, however, brought disillusionment. Only in one instance did he find the same male occupying the same web, and he doubted whether even this spider had been as constant as he appeared. 'I could not help suspecting,' he comments, 'that the one apparently faithful male had paid visits to adjoining webs in the interval and returned thereafter, quite by chance, to the same female.'

The actual process of mating in spiders is unique. The male spins a very small web and then stimulates the ejection of a drop of sperm by jerking his abdomen against it. The sperm drops on the web and the spider absorbs it into receptacles, like fountain pen fillers, in his palps by dipping them into it and then, in some species, holding them up while the sperm soaks in. This process of 'sperm-induction' may take anything from three to thirty minutes according to the species.

When sperm-induction is complete, the male goes in search of the female and mating consists of the insertion of the male palps into the body of the female and the discharge of their contents. When this is accomplished, he leaves her as quickly as may be. In most spiders, it is necessary to carry out sperm-induction all over again after each act of mating, but this is not invariable.

3

THE MOTHER

IT is hard to deny that the female spider has certain shortcomings as a mate, but she shows up rather better in her relations with her young. Even here, however, caution must be exercised in order to avoid being deceived into forming a higher opinion of her maternal behaviour than the facts justify.

One of the hunting spiders, for instance, weaves an egg-sac of the finest silk and deposits her eggs in it. Her task is far from finished for she now attaches this cumbersome bundle, which is nearly the size of a cherry, to her own body and for several weeks—until the eggs hatch, in fact—she goes hunting and carries out all her daily business with her unhatched family bumping at her heels. By night or by day she never deserts it for a moment and will defend it with brilliant courage even against insects which would normally put her to flight. 'If I try to take the bag from her,' writes Fabre, 'she presses it to her breast in despair, hangs on to my pincers, bites them with her poison fangs. I can hear the daggers grating on the steel.'

He then made the experiment of tearing away her

egg-sac, in spite of her frenzied resistance, and giving her in exchange an egg-sac taken from another spider of the same species. The spider was instantly reassured, seized hold of the alien egg-sac and attached it to her exactly as though it had been her own. She then went off with it, apparently without remarking any change in her family.

Fabre argued, however, that the egg-sacs were very much alike and he decided that he had, perhaps, tried her too high in expecting her to know the difference. In consequence he devised another test for her. This time he gave her a ball of cork in exchange for her egg-sac. It had been formed into roughly the same size and shape as the egg-sac, but the texture must have been vastly different. Once more the change was unremarked by the spider who placidly attached this new object to her body as carefully as though it had contained her own precious eggs. Even when Fabre offered her a choice between a number of cork balls and her own bag of eggs, she made her selection quite at random, and simply seized whichever lay nearest. Her maternal devotion, in fact, may be striking, but it seems singularly ill-directed.

It might be supposed that it would reach a higher plane when the eggs were hatched and small spiders were in question, but further experiments carried out by Fabre scarcely suggest that this is so. The young spiders climb on their mother's back as soon as they have escaped from the egg and there they sit, maintaining their balance with considerable

agility during the whole of the next week or two.
Fabre remarks, 'Nowhere can we hope to see a
more edifying spectacle than that of the *Lycosa*
clothed in her young.' And indeed her fortitude
in carrying her family on her back for such a considerable period, in turning herself, as it were, into
an animated charabanc, does seem to argue an
almost unbelievable devotion and self-sacrifice.
Naturalists, however, are rightly suspicious of
'anthropomorphism' and Fabre's next experiment
seems to justify their scepticism.

He took a hair-pencil and swept the young spiders
from the back of one mother in such a way that
they fell close to another, who was also transporting
her family on her back. The children ran busily
about until they came across the legs of the rival
mother. Apparently without noticing anything
strange or new about them, they immediately ran
up them and established themselves on the back
of this already burdened mother, who docilely
accepted this considerable addition to her family.

Mother-love, it seems, will be extended to any
young spider. Filial love is easily transferred.
Fabre relates an incident which demonstrates this
fact under far more gruesome circumstances. One
morning he found two of his mothers engaged in
desperate battle. One of them was lying pinned on
her back while the other clutched her with her legs.
They were threatening each other with open jaws,
ready to bite, yet hesitating to do so. Finally the
victor darted her head forward and sank her teeth

into the quivering spider beneath her. The murder accomplished, she proceeded to make a meal of her victim, as Fabre says, 'by small mouthfuls.'

It might be thought that the children of the defeated spider would find themselves in an awkward situation, thus deprived of their support in life. They solved their difficulties with impressive ease. While their mother was literally disappearing before their eyes in this distinctly painful manner, they seized the opportunity of climbing on to the back of the victorious spider who was still occupied with her meal. She showed no resentment. She had slaughtered and devoured the mother, but she uncomplainingly adopted the orphans as her own.

A spider's devotion to her young may seem fanatical in many ways, but when once it is finished, it is gone without trace. The processes of her physiology bring her maternal urge temporarily to an end, and there is then no gentleness or forbearance left in her attitude to her children. The day comes when the young spiders climb down from the back of the mother who has carried them for so long. They scurry away and there is reason for their haste. If they stay too long they will have stayed too late, for the instincts of motherhood are nearly dead and a small creature, even though it is her own offspring, now suggests to her nothing more than a tasty morsel. The young spider who loiters may be eaten.

They know their business, however, and they begin running up blades of grass and the stalks of flowers. They are waiting to be carried upwards, for nearly all young spiders begin their independent career by becoming airborne. They choose a still, hot day for this migration so that the air rising from the warm ground may carry them into the sky. Their journey through the air may be long— Martin Lister saw them in the air above him from the top of York Minster in the seventeenth century, and Darwin noticed a shower of small spiders descending on his ship, the *Beagle*, when over two hundred miles from the nearest land. They are carried to great altitudes and aeroplanes have quite frequently run into the gossamer which they trail with them. Eventually they float back to earth, perhaps far away from their starting-point, and the cycle of mating and procreation begins over again in this new environment. This journeying through the air must have been largely responsible for spreading spiders over the world, and it also prevents overcrowding and the possible failure of the food supply which might result if spiders became too concentrated in one small area. It has the further advantage of removing them to a safe distance from their mother at a time when the instincts of motherhood have become temporarily exhausted.

Instinct in the world of spiders is unmodified by feeling or emotion, and is perhaps the more powerful because of it. There is the will to live, the will to mate, the will to care for the resulting young; but these seem to be generalised urges, turned on and off with disconcerting abruptness and having little relevance to any individual. The will to mate is strong in the female spider—she is a passionate creature—but the minute she is satisfied, she seems to become conscious of the need for food. She looks round, as one may surmise, notes that there is a succulent-looking creature just beside her, and thinks, 'A meal!' At any rate, whether she thinks or not, she is quite liable to snap him up, untroubled, it seems, by any recollection of their mutual pleasure of only a few seconds before. Sex has been switched off, so to say. Hunger is now in the ascendant.

The will to live is the most constant urge of all, but even that may be overborne by the violence of the urge to care for the family. A spider cannot distinguish her young from those of another, but she is none the less ready to die for them. There are some spiders which stand guard over their eggs in silken chambers and there at last they die from starvation, still faithful to their charge. Another kind devotes herself to sitting on her egg-sac in the open, defiantly protecting it from even the most formidable attackers. She never leaves her precious eggs for one moment, until at last they hatch out and the young spiders go their way. The mother is left alone. Her work is finished and now, very quietly, she dies.

4

ENEMIES

SPIDERS need favourable conditions if they are to
survive, for their enemies are numerous and
very terrible, and even the adult spider has much
to fear. Birds, fish, and toads will eat them and
insects will destroy them by means even more
horrible.

There is the larva of the ichneumon fly, for
instance. A pale cream larva can sometimes be
observed on the backs of small spiders. These
seem to be little disturbed by the unwanted guest,
though its sharp, hooked little teeth have pierced
their skin. They pursue the common activities of
their daily life—they spin their webs, they devour
their prey, they may even mate and lay eggs just
as though they were healthy and free. But their
doom is upon them, for the larva grows. This
increase is not sudden or dramatic—it is a steady
gaining in strength and size which is sustained by
the vital juices the larva sucks from the spider.
When the larva is nearly full-grown, the spider
for the first time shows visible signs of perturbation. She may make sudden and aimless rushes
about her web. She may neglect to repair the web

itself, or form it imperfectly. She may seem indifferent to the insects caught in her snare. The end is near. The larva, now almost at the peak of its development, murders its host by some means not yet discovered and makes a final meal off the corpse.

Certain hunting wasps serve the spider even worse, for, after stinging and paralysing her, the wasp drags her away to a burrow already prepared. She then lays her egg inside the spider, closes up the entrance to the burrow and goes her way, leaving her victim to endure her living death, the slow draining of her forces, until the end comes at last when the larva is almost replete.

The larva, on its side, guided by the promptings of a beneficent providence, is careful not to devour the vital portions of its living meal until the very last moment, for if the spider died too soon, its meat would go bad. The larva, however, arranges its gruesome timetable with commendable accuracy. It is not surprising that spiders react with an exhibition of hysterical panic when placed at close quarters with one of these wasps. Their legs become limp, they seem to sink to the ground and await with folded hands, as it were, the sting which is going to bring them to this terrible end.

Spiders are also the prey of *Acrocerid* flies. These lay their eggs amongst low herbs and the larvae which emerge from them leap on to the back of any passing spider. If the spider should prove to be of an unsuitable size, or should even turn out not to be a spider at all, the larva presumably jumps off and tries again. If it has judged well, however, and its host is considered satisfactory, it clings on by the hooks with which it has been conveniently furnished and creeps inside the spider's body, which it punctures with a special piercing organ. It spends the winter comfortably enough in the interior of its host, who, in her turn, seems to notice nothing seriously amiss. Spring sees the end of this union. The larva, which seems to possess as fine a sense of timing as the ichneumon and wasp larvae, leaves its host's vital organs as a titbit to celebrate the very end of its stay, and it emerges at last, full-fed and safe from retaliation, leaving the spider nothing but an empty shell drained dry.

Compared to these horrors, the hazards a spider faces from the attentions of birds, reptiles and even man seem comparatively mild, though they may cause more loss of life. She must guard against her own species, too, for one spider may prey upon another, and in China spider-fighting is a recognised sport. Dr. Bristowe dates his own interest in spiders from his accidental discovery, at the age of four, that two spiders placed together in a jam jar can be persuaded, by the judicious application of a thick grass stem, to fight to the death. His best fights were obtained when he matched the powerful but comparatively slow-moving *Dysdera* against the smaller but agile *Drassodes lapidosus*. Both are short-sighted and both aggressive.

Dysdera, an intimidating sight with her cream-coloured abdomen and brick-red thorax and legs, would pursue the smaller *Drassodes* relentlessly with formidable jaws wide open, ready to inflict a bite which would almost certainly be fatal if she could once get it home. Agile *Drassodes*, on the other hand, would walk about cautiously with her sensitive front legs groping in front of her. The first touch would signal the nature of her adversary and she would react instantly. She would arch her body and then, trailing a band of silk behind her, she would spring over and under and across her opponent's body at lightning speed, each time leaving her swathe of silk behind her and giving a passing bite as she went. *Dysdera*, hampered by the sticky silk threads thus spread about her, and wounded and weakened by the multiple small bites, would lumber ponderously after the elusive *Drassodes*. For *Drassodes must* be elusive or she is finished. If once those heavy jaws close in her flesh, the wound will be serious, she will lose her speed and agility and the fight will go against her. Bristowe writes that such battles may continue for an hour or more before one spider finally overpowers the other.

Man, too, must be reckoned a formidable enemy of spiders. If one spider is to be found to every three square inches of grassland, he must kill a large number quite unknowingly simply by treading on them or sitting on them. He does not kill so many on purpose as might otherwise be the case since they are protected by superstition. In Egypt they are thought to be so lucky that they are some-time placed in the marriage bed, and in England, this rhyme is still commonly heard:

'If you wish to live and thrive
Let a spider run alive.'

The general belief that it is unlucky to kill a spider may be at any rate partly responsible for a common inclination to put them out of the window in preference to squashing them, although the mess made by a squashed spider may be something to do with it, too. They are not left in peace, how-ever, if they attempt to live indoors, for their webs, at any rate, will be ruthlessly swept away by the ordinarily conscientious housewife. This caused their ever-faithful supporter, Dr. Muffet, consider-able distress.

'Surely miserable was her condition and estate,' he exclaimed, 'which in all that abundance and wealth, she only being indigent and bare, might not yet be admitted tenant on some short term of time, in some small odde corner, nor yet find one hole to live at peace in.'

In many parts of the world, and in England until fairly recently, spiders have suffered from the fact that they are regarded as an article of diet. Dr. Bristowe has sampled them and records that toasted lightly and dipped in salt they are, 'far from un-pleasant,' and taste something like raw potato mixed with lettuce. Large spiders may be placed on a skewer and roasted over a flame. Sliced up with chillies or eaten with salt, he declares that they remind him of a well-known brand of chicken essence.

Dr. Muffet has something to say about this practice, too, and records, a little sadly, that, 'There was one Henry Lilgrave, living not many years since, being Clerk of the Kitchen to the Right Noble Ambrose Dudley, Earl of Warwick, who would search every corner for spiders, and if a man had brought him thirty or forty at one time, he would have eaten them all up very greedily, such was his desirous longing after them.'

Not many people 'long after them' today in this country, but until quite recently, even if they were not eaten for pleasure, they might still be taken for the sake of the health. Indeed, spiders make a considerable showing in old prescriptions. Eaten in handfuls on bread and butter, they were said to be an effective remedy for constipation. Rolled up in butter and swallowed as a pill they were thought to prevail against jaundice. For malaria, it was generally considered necessary to swallow the creature alive and wriggling and without disguise.

Spiders no longer figure in prescriptions, and faith in their power to bring good luck, if left unmolested, or ruin and disaster if killed, no longer has the same force as it used to in days that are past. Spiders no longer even command respect as prophets of the weather. The spider is left with one solitary claim on our forbearance, but it is an important one. As has been said before, they are the unrivalled destroyers of insect pests. In England and Wales alone spiders consume not less than 220,000,000,000,000 insects in one year.

But a debt of gratitude will never suffice to inspire love. If the mere appearance of a spider gives rise to horror and disgust, an inquiry into their lives and habits does little to induce a gentler regard. They are savage assassins, slayers of their mates; devoted to their children, it is true, but with a crude, instinctive violence that owes nothing to affection. If they are industrious, it is only that they may eat. If they save us from being plagued by flies, there is no altruism in the act. They share the bitter, savage struggle for existence which prevails among the insects on which they prey.

BEETLES

1

THE FEAST

'*C*oleoptera for men!' exclaimed Oliver Wendell
Holmes in the person of his character 'the
Scarab'; but in spite of this pronouncement, few
people seem to have made beetles their favourites.
Fabre is the one outstanding exception and he seems
to have been chiefly attracted to the Dung Beetle
by the high moral qualities he claimed to discern
in this unpromising creature.

A hasty judgment is too often unjust and Fabre
no doubt felt it would be wrong to condemn the
admirable corporation of Dung Beetles simply
because their choice of nutriment seems, to our
taste, unfortunate. As Dr. Muffet, that ardent
lover of insects, says, the fact that the beetle 'useth
the excrements of living creatures for its own com-
modity is no fault, but a commendation of its wit
and ingenuity.' Besides, more careful consideration makes it obvious that scavengers are a necessity
in the interests of health and hygiene, and if
scavengers are attracted to their unsavoury business
and embrace it with enthusiasm, so much the better
for the rest of us. The Dung Beetles, in particular,
perform their necessary tasks with ingenuity and

skill and appear to bring to them a fund of industry,
zeal and good humour which we may well admire.

Their food is not difficult to come by. Sheep,
horses, cows all unite in distributing, as Fabre puts
it, 'manna to the enraptured Dung Beetles.' The
smell of fresh dung attracts them from a considerable
distance and they come hurrying from all directions with a rapidity born of greed and the certain
knowledge that others, too, will desire their share
of the feast. The Scarab Beetle, sacred to Ra, is
sure to be among them, a large and impressive
creature in shining black, lumbering along with
an air of anxious preoccupation. His size gives him
an advantage and there is no time for good manners
with so many hungry mouths snatching at the food.
He pushes his way to the front, knocking over one
or two of the smaller and less well-endowed beetles
as he goes.

The hungry beetles are of several different sorts
and do not cherish the same ambitions. The
smallest hope to snatch what they can, when they
can. They hastily swallow any little crumb that
may come their way—instinct must warn them
that their chances of getting clear away with a
desirable portion are too small to be worth con-
sidering. Others plunge into the middle of the
heap, secure their helping and bury themselves
with it directly below ground. The majority, however, appear to feel that they cannot properly
savour their meal in the midst of a jostling throng.
Their aim is to secure a good portion—enough for

two or three days of uninterrupted eating—and get it away to some solitary retreat where it can be enjoyed at leisure and without anxiety.

The Sacred Beetle is among them. His first care is to select his portion, and this is by no means made up of what first comes to hand. He is in a position to choose and he does so, selecting and rejecting with the arrogance of strength. Lesser beetles may make what they will of the less nutritious vegetable fibres—they are not for him. A worthy portion is thus selected from the smoking pile and transport becomes the next problem.

Instinct—or the harsh goad of evolution—has taught him that the maximum amount of dung can be moved across country with the minimum of effort if it is first moulded into a sphere. In the course of long centuries of time the right equipment for such an enterprise has been developed, and the busy beetle plies his tools with concentration and efficiency. His strong, bow-shaped front legs, well-equipped with teeth, collect the material together and push it beneath him. The four hind legs are not so strong, but they are long and thin and bow-shaped and seem perfectly designed for the next stage of the work. They press the material together and set it spinning until at last it is rounded and firm—the perfect sphere which is needed for the business of a move across country.

The beetle now harnesses himself to his burden— that is to say, he rears himself on end, places his hind legs round the ball of dung and lets his middle and front legs rest upon the ground. His strong front legs supply the motive force. He presses them into the ground and soon the cumbersome ball begins to move, slowly and then faster until at last, in this surely somewhat strained position, the beetle begins his journey.

It is not an uneventful journey. It seems a strange oversight on the part of nature that this hard-working beetle, so adept in forming his ball of dung, so enthusiastic, it appears, and so resolute in undertaking an enterprise that must, at the very best, be extremely arduous, is yet so lacking in elementary prudence that he sets off on his journey travelling backwards with his cumbersome piece of luggage bumping along in the front. It is true that by this means he can use his strong front legs to the best advantage. Other advantages might be adduced with a little thought. And yet, in the end, nothing seems to compensate for the one insuperable handicap under which the beetle struggles. There can be no question of planning a route. There can be no possibility even of avoiding obstacles or sudden disaster. The beetle is obliged to follow a sporting course and meet such mishaps as come his way with as much philosophy as he may be supposed to possess.

In the first place, the terrain can hardly be expected to be entirely flat. Sooner or later a slope must be encountered and it may lie at an angle to his path. The problem of pushing a ball of dung somewhat larger than himself up such an obstacle can scarcely be simple. The beetle is not prompted, however, to take the easy path downhill. He strains stubbornly onwards, but the success he surely deserves is seldom won at the first attempt. His foot may slip momentarily on a smooth piece of gravel—it is enough to bring disaster. Down the ball will roll to the bottom of the slope, flinging the beetle on his back. For the moment, he lies there with his legs helplessly waving.

He is not downcast, however. His demeanour, at any rate, suggests unshakable determination. He struggles to his feet, runs down to his ball and gets in position to push once more. Still he makes no concessions to the lie of the land, the downward path does not tempt him. Up the slope with all its hazards he struggles again, until perhaps a grass root this time, or some other obstacle, blocks his progress. He does not give in. It is impracticable to back down and attempt to steer round the obstacle. At any rate, the idea does not appear to occur to him and the foolish, headstrong creature continues to strain and heave with little effect, until at last the ball is forced slightly to the side and once again the slope takes it. Perhaps it takes the insect too this time, and both go hurtling once more to the bottom. But again he is not dismayed and scarcely pauses to collect himself before he is once more at his task.

And in the end, somehow or other, he manages to proceed on his way. One would imagine that a prodigious appetite must be developed in the course of this trying journey. The trials, too, are not confined to the contours of the ground. The world of beetles appears to be no more immune than is our own from the threat presented by lazy, worthless creatures who are too indolent to secure a portion of the world's goods for themselves, but hope to profit from the labour of more conscientious citizens.

Another beetle, who has taken no share in the labour of shaping the ball, may form the design of sharing in the feast to come. He rushes upon the first beetle—who, owing to his position in relation to the ball, is handicapped in the matter of self-defence—and sends him flying. The attacker then adopts a position of considerable strategic advantage—he balances himself on the top of the ball. The enraged owner is anxious to win back his property, but it is an affair of some difficulty. If he reaches up to strike at the robber, he is likely to receive a blow which will knock him off his legs. He circles cautiously round, searching for a favourable opening. His enemy circles with him, prudently facing in the direction from which attack may come.

The situation seems hopeless. Cunning is the only resource left for the dispossessed beetle. He turns his attention to the base of the ball and presses against it until at first slowly, and now faster, it begins to move. The position of the thief is no longer so secure. Faster and faster moves the ball and the beetle on the top can only keep his balance by the most violent exertions. It cannot go on for long. Sooner or later the ball bumps over a pebble and the brigand beetle falls to the ground.

And now indeed battle is joined. Breast to breast they struggle and heave, their legs grapple and intertwine, there is a grating crash as armour meets with armour. Yet in the end these encounters are singularly bloodless. One or the other grows tired of the struggle and decides that a meal can be more easily won by conventional means. He gives up and repairs instead to the diminishing heap of dung to form another ball, while the beetle in possession goes on with the journey.

Equally nefarious intentions, however, may be concealed by a friendly demeanour, and a rival beetle may not always declare himself an enemy in this unmistakable fashion. He may present himself, perhaps, in the guise of a 'partner'—a partner, however, who wishes to share in the profits rather than the ardours of their mutual enterprise. He is generally received without question or resentment by the beetle in possession and in the early stages of their relationship the newcomer seems anxious to repay this good feeling. He displays, at any rate, a token anxiety to be of help. The rightful owner is in the position of honour at the back of the ball, so the new assistant places himself in front. He takes the ball between his front legs and proceeds backwards on his hind legs, pulling the ball towards him.

Everything seems satisfactorily arranged. The

uninvited assistant is not, perhaps, doing a great deal to deserve his dinner. Since he is going backwards and can see his way no better than the owner, his presence does not even help to avert the constant accidents and tumbles which attend the journey. Still, these are endured with resignation and apparent good humour, and at least the 'partner' has made some gesture of co-operation.

Even among these unsatisfactory partners, however, some sink to depths of depravity which would be shunned by others. After this initial display of good-will the unwanted assistant sometimes appears to feel that he has done all that can reasonably be required of him. At any rate, he proposes to do no more and he flattens himself on the ball and allows the lawful owner to roll him along with it. It seems that he now expects a free ride as well as a free meal. The harassed owner, uncomplaining or unperceiving, makes no protest, but labours on at his task until some inner voice of thought or instinct warns him that the time has come to desist. The journey is over, the hour of feasting is at hand. But first the banqueting-hall must be dug.

It would be pleasant to record that even at this late hour the dishonest beetle shakes off his lethargy and sets to work with a will. This may indeed happen, but, if so, it has never been observed or recorded. Fabre, who almost certainly knew more of the habits of this beetle than any other naturalist, takes a pessimistic view of his probable course of action. The true owner chooses his spot and begins to dig with a fine display of energy. Soil flies out behind him and his head begins to disappear in the ground. Soon his body disappears, too, and the second beetle, apparently asleep, or at the least unaware that work is going ahead in which he might well be of considerable assistance, remains motionless on the ball.

Nobody can know whether the rightful owner now begins to feel the first stirrings of suspicion. It is known, however, that he returns to the surface at fairly frequent intervals and, whenever he does so, he looks first to the ball, apparently to make sure that all is well. Sometimes he even moves the ball nearer to the entrance of the hole before going below again. The other beetle still makes no move, while in the ground beneath the banqueting-hall grows ever larger and of nobler proportions. Soon, indeed, the excavations are so vast that it is no longer practicable for the working beetle to make more than very occasional appearances on the surface.

It is now that this treacherous partner reveals his true character. He awakes abruptly from his convenient sleep, harnesses himself to the ball of dung and for the first time shows the vigour of which he is really capable. He rolls the ball away with a fine turn of speed. He may have travelled some distance by the time the true owner returns to the surface, but generally the smell of dung reveals where he has gone and the pursuit is on. The malefactor is generally caught—the unencumbered beetle has all the advantages in running across country—but he resourcefully tries to conceal what he has been about. He quickly reverses his position, seizes hold of the ball with his front legs and appears to be holding it as though it had rolled off down the slope and he had conscientiously hastened to stop it in its course. According to Fabre, the first beetle seems to accept this explanation of events. At any rate, he shows impeccable good nature, makes no smallest attempt to attack his shifty partner, but accepts his help in bringing the ball back to the burrow. And when the burrow is completed, he also accepts his presence at the meal which he has done so little to secure. With irreproachable good manners the lawful owner even allows him to eat an equal share of the rich feast.

There must be times when the partner gets clear away with the valuable prize of dung, and even then the good nature and resignation of these beetles do not falter. The rightful owner has collected these provisions for his own enjoyment, he has brought them from far away and surmounted many difficulties on his laborious journey. He has dug a large banqueting-hall underground and at last the reward of all his effort seems to be at hand— the long feasting, the sumptuous repast. And yet, at the very moment when satisfaction seems in his grasp, it is snatched away. All is to be done again. This much-tried beetle could scarcely be blamed if he gave way to despair and abandoned further effort. According to Fabre, however, this is not so.

'The Dung Beetle does not allow himself to be cast down by this piece of ill-luck,' writes Fabre. 'He rubs his cheeks, spreads his antennae, sniffs the air and flies to the nearest heap to begin all over again. I admire and envy this cast of character.'

It seems none the less worthy of admiration even if we must attribute the beetle's behaviour not to choice, or thought, or to any conscious process whatsoever, but to the blind promptings of physiology at work with hormones or nervous impulses or some mechanism not yet understood. And next time, perhaps, all may go well. He may have the good fortune to transport his prize without interference and without attracting the attention of an uninvited collaborator. He may succeed in completing his underground chamber without losing his ball of dung to a sneak-thief, and finally lower it into position without mishap. Then he shuts up the entrance with a little debris which he has reserved especially for this purpose, and at last he is safe. He can give himself up without reserve to the pleasures of the table.

The imposing ball of dung almost fills the entire chamber. There is just room for the diner to tuck himself between his sumptuous store of food and the wall at his back. It is dark, but perhaps he may find this pleasant. The hot sun is excluded, there is nothing to disturb or distract. He can concentrate on the business in hand.

Day-long, night-long the feast goes on. The beetle may eat without a single pause, it is said, for

as much as a week or even a fortnight. His digestion is admirable and well-designed to solve what
might otherwise become a considerable problem.
He never pauses, he never stirs, but behind him
accumulates, in the form of an unbroken cord, the
end-products of his digestion, the unusable remnants of his gargantuan meal.

At last, when all is gone, the beetle emerges
from his underground orgy full-fed, and, as one
might suppose, satiated. But this is not so. Without
pause, without hesitation, he hurries off to the
nearest patch of dung to begin the cycle all over
again.

This simple life of work and feasting lasts only
for a month or two. In June or July, the heat becomes too fierce for such efforts and the Sacred
Beetles bury themselves underground for the rest
of the summer, to emerge in the autumn with
other preoccupations.

2

FAMILY CARES

IN autumn the Sacred Beetles must occupy themselves with the future of the race. The balls of dung which they had formed and rolled across country with so much address and perseverance in the spring were for their own delight. In autumn, the mother forms a ball which is to provide for her young. It is formed with more care than a ball which is merely for her own consumption—only the finest materials are used—and the final shaping is carried out in the burrow itself.

This ball is generally about as large as an orange, but when once inside the burrow its shape is transformed into that of a pear. The egg is laid at the narrow end, where a certain amount of air can penetrate—a necessity for the developing grub. The grub's food, on the other hand, is stored in the thick part where, protected by a hard outer shell, there is little danger of its becoming dried up and too hard for the grub to eat. As soon as the grub—the larva—hatches out of the egg, it begins eating and at the end of four or five weeks, now a fat and rather bloated creature, it sheds its skin and transforms itself into a pupa. Immobile, the colour of amber, it passes about another four weeks in this stage before the perfect beetle finally emerges.

The male Sacred Beetle may occasionally give some small assistance to the female in her labours for her children, but, if so, it has never been observed. Like other males of the insect world, he is in the habit of treating his family responsibilities with regrettable unconcern. When once his necessary part in the business of ensuring the future of the race is performed, he shows no care for the consequences of his pleasure. The mother is left to wear herself out for her family unaided. For his part, he is more interested in providing regular and lavish meals for himself, or in paying his court to females less preoccupied with serious domestic concerns.

The little *Sisyphus* Beetle, one of the smallest and most hard-working members of the family of Dung Beetles, provides a welcome exception. He proves himself a model husband and an excellent father, a thing almost unheard of in the insect world.

The union of the male and female of this insect is consummated among the sheep dung on which they have lately fed. But this is no passing affair, the passion of a moment. The male demonstrates from the first his intention of sharing in all the anxieties and effort attendant upon the founding of a family. He remains with his mate. He assists her to form the little ball of dung which is to provide for their children to come. He is her partner in the rough journey across country to a suitable site for the burrow. The mother takes her place in front and walks backwards, dragging the ball after her. The father pushes from the back, adopting the same inconvenient position as the Sacred Beetle—fore-legs planted on the ground, tail in the air, back legs clasping the ball. It is quite impossible that either should see where they are going, but they set off with a will. Whether indeed they have some destination in mind or whether their direction is dictated by the whim of the moment is a question that nobody has yet been able to answer.

The journey is adventurous, as might be foreseen, but the two beetles are not dismayed by the hazardous course which chance obliges them to follow. They stumble, they tumble, they are flung violently on their backs and go bouncing off down slopes. It is impossible to discern the very smallest sign of ill-temper, or even of impatience. They 'pick themselves up, dust themselves off and start all over again.' For hours and hours they persevere in this unreasoning determination to reach some unknown goal. At last, whether from exhaustion or because some inner prompting tells the mother that here and here alone will her family grow and thrive, they come to a stop.

They have still not settled on the exact spot for the excavations, however, and the mother goes off to investigate. The father is left in charge of their ball and sometimes whiles away the time of waiting by spinning it between his legs.

Seeing him frisking in this joyful occupation (remarks Fabre), who can doubt that he experiences all the satisfaction of a father assured of the future of his family? It is I, he seems to say, who have made this loaf so beautifully round; it is I who have made the hard crust to preserve the soft dough; it is I who have baked it for my sons! And he raises on high, in the sight of all, this magnificent testimonial of his labours.

As soon as the mother has chosen the place in which the hole is to be dug, they roll the ball together to the site and the mother begins operations. All the time she digs, the ball is close behind her— the mere feel of it appears to inspire her to fresh efforts—and the father, her conscientious partner, keeps a firm hold upon it. Soon she disappears below ground and the ball is carefully lowered after her.

At last, the burrow is big enough to accommodate the whole of the ball and the parents as well, and the party disappears from sight. No doubt the finishing touches are being given to the apartment before the egg is confided to the ball of dung. According to Fabre, as much as half a day may be spent at these tasks. Sooner or later, however, the father reappears, but not in order that he may now go away on his own concerns. He has decided, it appears, that he can only be in the way while his wife carries out the peculiarly feminine duties which now detain her. Or, if this explanation is too fanciful, it may be that there is simply no room for him below while the female is occupied in laying her egg. At all events, he makes himself as comfortable as he can near the mouth of the burrow and settles himself down to wait. Clearly, it is beyond his competence to take any part in the laying of an egg, but at least it can be claimed for him that he has stood on guard and afforded his mate the moral support of his continued presence.

The female beetle reappears after about a day, and at once the faithful pair set off in search of more dung. A satisfactory family cannot be achieved on the strength of one such enterprise. The whole operation must be repeated many times before their aspirations are satisfied—or their physiological urges exhausted.

It seems a charming picture of loyalty and mutual endeavour. Fabre, however, is not prepared to assert that all *Sisyphus* Beetles attain this high moral standard in their behaviour. 'There must,' he says, 'be flighty individuals who, in the confusion under a large cake of droppings, forget the fair confectioners for whom they have worked as journeymen and devote themselves to the services of others encountered by chance; there must be temporary unions, and divorces after the burial of a single pellet. No matter: the little I myself have seen gives me a high opinion of the domestic morals of the *Sisyphus*.'

Their fertility is also superior to that of the Sacred Beetle and for this, as Fabre maintains, there can be only one explanation. The family must always thrive when two bear the burden of its support and the mother is not left to struggle alone.

If the *Sisyphus* Beetle may justly lay claim to the proud position of the most conscientious father in the whole of the insect world, the female of the *Copris* Beetle—yet another Dung Beetle—displays virtues almost equally rare as a mother. The two sexes of this beetle labour together to build the underground burrow which is to serve alike as nuptial chamber and nursery. Together they stock it with provisions for the coming generation; but, for the father, that is the end of labour. The mother now begins a solitary retirement which is to last for four months. She settles down in the midst of plenty, but she does not eat. This selfless insect will not take one mouthful of her children's portion. She waits, starving and alone, throughout the whole of the four months, occupying her time by caring for her balls of dung and their growing occupants. She feels them with her antennae. She repairs any little imperfection she may note, and smoothes and polishes their surface. Sometimes she seems to be listening to the sound of the tiny grubs moving about inside.

The rains of October soften the shells of the balls of dung, and the young *Copris* Beetles, now fully-formed, are ready to emerge. They break their way out and soon this exemplary family, devoted mother and healthy brood, emerge into the daylight. For the first time after four months of starvation, the mother may delight herself in feasting—with the added satisfaction, as one might suppose, of seeing her young about her.

It would be pleasant to record that even now her maternal care does not falter or fail, but the truth must be faced. Abruptly and at once, the very minute she comes to the surface and feels the sunlight on her back, her family is forgotten. She bears them no love, she shows them no care. To be entirely frank, she seems incapable of recognising them. In the hard adult world, each insect must now fend for itself.

3

SEX AND SLAUGHTER

THE Dung Beetles, humble and useful scavengers, afford a pleasant impression in their lives of loyalty and a sense of duty. Even though they are, in fact, driven by blind urges and no conscious thought ever troubles what little brain they may be said to possess, still the impression persists, and it is undeniable that their actions do seem to express these desirable qualities. A brother beetle, the savage Golden Gardener, is lamentably different. These beetles are chiefly remarkable for the dispatch with which they are able to slaughter and consume a very large number of victims. Fabre compares them to the celebrated meat factories of Chicago into which, it is said, millions of live animals enter every year and rapidly emerge again in the form of preserved meat, sausages or rolled ham.

Fabre possessed twenty-five Golden Gardener beetles, or Golden *Scarabaei*, as he called them, in his insectary. (They were probably Golden *Carabus* beetles.) When no food was at hand, it was their habit to lie dozing under a piece of board which provided shelter from the sun. Caterpillars are their favourite food and by good fortune Fabre came upon a procession of these creatures one day, descending from a tree. He decided that they would provide a good flock to offer up to his Golden Beetles.

He captured these Pine-Processionary caterpillars and placed them in the insectary. There were, perhaps, one hundred and fifty of them. They appeared to be little incommoded by their abrupt removal, re-formed themselves into line and began to move forward in rhythmic undulations across the sand of the insectary. By chance, their course passed fairly near to the board which concealed the sleeping murderers. 'I cry Havoc!' writes Fabre, 'and let loose the dogs of war. That is to say, I remove the plank.'

First one sleeper then another began to stir as the scent of prey penetrated his senses. One of them ran towards the caterpillars. Two or three others were close behind. In a minute, the whole pack of them had rushed to the attack and the work of slaughter had begun. Wounds were soon torn in the furry skins of the caterpillars and their bright green entrails were shed upon the ground.

In vain the wretched creatures writhed and wriggled, in vain they dug their feet into the sand and strove to bite the savage beetles. They could do nothing. Those who were still whole began desperately to dig in the hope that they might hide themselves beneath the sand. The beetles frustrated their attempt. While they were still half out of the ground, they were seized and killed. The beetles seemed to run everywhere at once, biting, mutilating, murdering.

A few minutes was all that was needed for the destruction and partial consumption of one hundred and fifty plump caterpillars—a remarkable piece of work. Soon nothing was left but a 'few shreds of still palpitating flesh.' The twenty-five Gardener Beetles ran busily about on the field of slaughter, tearing and tugging at these remnants, munching as they went.

'Only the ear of the mind could bear the shrieks and lamentations of the eviscerated victims,' remarks Fabre. 'For myself, I possess this ear and I am full of remorse for having provoked such sufferings.'

However, he was soon busy provoking some more. The intestine, he reflected, rules the world. Until we can become independent of the stomach and its demands—a desideratum unhappily not yet attained—we must bend our energies to satisfying them. For us, the meat canneries of Chicago; for the Golden *Carabi*, the caterpillar. And since their taste is catholic and their hunger imperative, the variety of caterpillar is of little importance— they are prepared to eat any caterpillar that offers, provided only that it be of a convenient size. If it is too small, the beetle seems to feel that it will not supply a mouthful worth the trouble of catching. If it is too large, the beetle is sometimes unable to deal with the situation. An attack merely provokes a violent contortion on the part of the proposed victim and the would-be murderer is flung on his back.

Provided, however, that the size of the caterpillar falls within certain limits, the Golden Beetles do not complain of variations in taste or texture. The Pine-Processionary caterpillars are equipped with a corrosive poison—as Fabre remarks, they must form a somewhat highly-seasoned dish—yet they are devoured with apparent enjoyment. Other caterpillars are covered with a thick furry coat, but the Golden Beetles admit them to their diet without demur. Fabre found that one caterpillar, however, did cause a little difficulty.

He decided to see if his Golden Beetles could deal with the Hedgehog Caterpillar of the Tiger Moth. This is the hairiest of all French caterpillars and is an altogether impressive insect with its striking mane of red and black. Fabre introduced it to the company of his band of assassins.

At first, nothing happened. Occasionally a Golden *Carabus* would reconnoitre the imposing stranger with a slightly wistful air. He might even essay a bite, but would retire at once without doing any damage, evidently unable to make any impression on the thick mass of hair. He would come back, however, and continue his puzzled inspection. The caterpillar, for her part, received these overtures with an air of slightly scandalised dignity.

The stalemate lasted for some time, but it could scarcely be permanent. Hunger gives courage and sharpens the wits. It even seems to have given birth to an alliance on this occasion, for the Golden Beetles seemed to come to the conclusion that where one had failed, a number might succeed. At all events, whether by conscious design or by chance, four beetles made a concerted attack on the Hedgehog Caterpillar.

Aloof, majestic, impregnable, as she might well think, the caterpillar went undulating on her way, seeming scarcely aware of the four beetles scurrying round her. Suddenly they flung themselves upon her. As usual, her thick coating of fur seemed to choke them; but this time they persevered in the attack. They rushed on her again and again, gnashing their mandibles and tearing at her hairy covering. She was beset in front and behind and at the side. The battle could not long be in doubt. In the end, she succumbed as easily as the merest grub and was devoured without trace.

Life in the insect world is red in tooth and claw. Such encounters multiply themselves endlessly. But at least the insect may plead that he kills that he may live, that hunger and not malice is his goad. The Golden Beetles, however, cannot point to an exemplary domestic life such as that of the *Sisyphus* Beetle in extenuation of their savage slaughter. Even in his dealings with the opposite sex the Golden *Carabus* gives little sign of gentleness.

He approaches his proposed partner in the midst of a crowd and wastes no time on preliminary overtures. With an air of businesslike efficiency, he flings himself upon the female without warning.

She, for her part, shows no sign of perturbation or shock, but simply raises her head a little in what may be taken as a gesture of consent. The male then 'beats the back of her neck with his antennae,' and, after this somewhat rude gesture of amorous interest, the embrace proceeds. It is brief and the parting is as abrupt as the meeting. The pair pause only for a short interval of refreshment before going in search of fresh adventure.

There can be little sentiment in such hurried couplings, but we should hesitate to criticise the Golden *Carabus* too harshly. There is something to be said for a friendly and practical spirit in these matters, and sentiment is not always an unmixed blessing. This the Golden *Carabus* may know to his cost, for in one matter the male appears to be invincibly sentimental.

By the end of June, the female's interest in sex has exhausted itself. The male no longer appears desirable to her as a mate. Unfortunately for him, however, he now exercises attractions of a somewhat different nature. If a female encounters a male in open ground, she rushes at him at once, for he is nothing more to her now than an insect smaller than herself and well-suited to her table. She seizes him and pulls at him, gnawing at his flesh as she does so. The victim, who may be full-grown and healthy, apparently makes no effort to ward off the attack or launch one on his own account. He confines himself to pulling desperately away from her in an effort to escape and free himself from her jaws. The struggle surges to and fro as now the female, now the male, appears to be winning. Sometimes the male succeeds in his efforts and gets clear away. More often, however, the savage female has her way with him. His skin at last gapes open, his entrails are torn out and eaten, and he dies, still without making any attempt at retaliation.

Here is a mystery. He is able to defend himself, he is strong and could, no doubt, acquit himself with credit, and yet he feels, as it would seem, an invincible repugnance towards 'raising his hand to a woman.' Can it be that he remembers the ardent embraces of earlier days? Does he regard her as the mother of children which may well be his? Nobody would seriously suggest that such flights of fancy could be the true explanation of his behaviour. And yet what are we to say of this chivalrous insect who will not defend himself against the females of his own kind and even sacrifices his life for such a scruple? Whatever the basis of his conduct, to most of us such chivalry would appear the grossest folly. But it may be that these small creatures experience raptures and ardours to which we human beings can scarcely aspire.

Fabre points out that the scorpion of Languedoc allows himself to be devoured by the female when the act of mating is finished. He is equipped with a deadly sting and could very well use it in his own defence; yet he does not do so. More striking still, the male of the Praying Mantis clings to his mate in ardent embrace and does not relax his grip or attempt to escape even when she proceeds to eat him up 'by small mouthfuls.'

Such extremes of restraint are too difficult for us to understand. What man is there who would be so carried away by the ecstasies of passion or the urgencies of lust that he would allow his partner to devour him by 'small mouthfuls'—or even by large? In such matters we must admit ourselves, by comparison with certain of the insects, cold-blooded and calculating.

Not all insects, however, possess this abandon. The females of some of the Oil Beetles, in particular, seem distinctly phlegmatic. The male *Cantharides* Beetle, for instance, one of the Oil Beetles, seeks out a mate who will almost certainly prove to be engaged upon a meal. Undeterred, he flings himself upon her, encircles her with his hind-legs and gives her a good hard slapping with his abdomen, first on one side and then the other. This vigorous advance leaves the female in little doubt as to his intentions, but he none the less follows it up by violently lashing her neck with his antennae and fore-legs. He appears to feel that a rousing approach is necessary if he is to get any response from his sluggish mate.

The female appears to bow to the storm, but the minute the blows slacken she resumes her interrupted meal without giving any sign of yielding to her suitor's importunities. The male hurls himself into another frenzy and the blows rain down once more. The female tries to lower her head, but he seizes her antennae with his fore-legs and forces them upwards. In this commanding position, like a rider sitting a mettlesome steed, he continues to batter and lash his chosen mate into a more compliant frame of mind.

At last she yields and the embrace takes place. It goes on for twenty hours. This considerable period of time appears to hang somewhat heavily on both partners, for the male makes several attempts to uncouple himself from the female, but without success. For her part, she seems to wish heartily that the whole thing might be over sooner, but philosophically accepts the inevitable and beguiles the time by eating. The deflated male is dragged behind her from leaf to leaf while she selects a tempting morsel, now here, now there. Sometimes he makes a gesture of independence and begins to eat something himself, but nothing can save his position from indignity, particularly when the female takes to her wings and drags him after her as so much dead weight.

It seems a pity that the female *Meloë* Beetle, another of the Oil Beetles, does not experience a keener satisfaction in mating, for its aftermath of motherhood brings her considerable labour. So great are the perils which threaten the young larvae of this species, so formidable are the hazards which they must overcome, that she is obliged to lay a prodigious number of eggs—obliged, that is, by the process of evolution. For if she laid fewer eggs, this species of beetle would assuredly die out.

The female *Meloë* Beetle produces about four thousand eggs in one batch. This might be thought enough to satisfy the most unbridled maternal instinct, but it does not satisfy the *Meloë* Beetle. She glues these eggs together and deposits them in a hole which she digs in the ground, and then proceeds to lay another batch. And another. Ten thousand eggs is quite a moderate family for the female *Meloë*.

The task of laying completed, and the eggs safely stored, the mother appears to feel that she has done all that the most exacting could demand of her, and she goes her way. This horde of children are left to fend for themselves in the struggle for life, and for them it is a struggle indeed.

The larvae hatch out in May or June. They are little louse-like creatures, pale yellow, flat and elongated, about the size of a semi-colon (;). After a brief period of torpor, they burst into frenzied activity and can be seen busily climbing up and down blades of grass. Their efforts are not as futile as might appear for they are looking for something. Their purpose, whether conscious or unconscious, is to find some low-growing plant, such as the celandine or primrose, and establish themselves in a flower whence they may embark on the next stage of their journey in life.

It is a scene of anxious haste. And no wonder! Nature has apparently endowed these little creatures with some botanical discrimination so that they may pick out a plant which is likely to be visited by a bee, for a female bee of a particular species is the next essential in their complicated phases of development. Nature has further endowed them with the instinct to attach themselves to a hairy leg; but here, it must be confessed, she has been a little remiss. One hairy leg, after all, is very like another, and flies as well as bees possess them, males as well as females. Moreover, Fabre carried out experiments which demonstrated that their instinct in this important matter was far from reliable.

He found the larvae in large numbers gathered on appropriate plants awaiting the arrival of a bee. He offered them a blade of grass, a piece of straw, the handle of his tweezers—all were accepted by the foolish larvae with equal enthusiasm. It is true that they became restless almost at once and made efforts to get back to the plant, if it was not too late, but the hard life of the wild seldom offers second chances. Fabre, however, did. He put them back on the plants and offered them the same objects again. This time they were more wary, showing, as he says, that these 'animated specks' have 'a memory, an experience of things.'

He decided to offer them objects rather more similar to their desired hairy legs. He offered them small pieces of velvet or cloth, fragments of cotton-wool, or fluff from the everlasting flowers growing near at hand. Again they were seized on by the larvae, but again they were rejected after the first moment or two—that is, the larvae quickly became restless and tried to escape back to the flower.

Fabre then offered the larvae live insects. He offered them drone flies, blue-bottles, hive-bees and small butterflies. Once even a large black spider. Here discrimination broke down completely. The larvae crowded on to these creatures and settled down with every appearance of satisfaction. There was no sign of restlessness, no smallest hint that the larvae were conscious of anything wrong, and no attempt to get back to the flower. And yet none of these insects can provide the home-background the larvae need in order to achieve their next stage of development.

No wonder the *Meloë* Beetle is so prolific! She needs to be.

It is unnecessary to linger over the unfortunate larvae who choose wrong at this crucial moment of their lives. They have no future. They have cast their dice and they have lost. They will starve slowly to death. It is in the hands of the others, the fortunate few who have managed to attach themselves to the right kind of bee and a female one at that, that the future of the *Meloë* race rests.

A Mason-bee hovers over a flower. It alights, the flower shakes and at once any resident larvae become furiously active. They run to the edges of the petals, they stretch out their legs in every direction, feeling, searching. If one of their legs comes into contact with the hairy leg of the bee, the larva

jumps nimbly aboard, takes up its position—generally on the thorax—and nestles down for the
journey.

The imperceptive bee, no doubt unaware that
she has now a passenger on board, surely unknowing
that she is carrying with her the doom of her family,
at last returns to her home. The larva becomes
alert again at once. It leaves the bee, settles down
in the bee's nest and at once proceeds to devour
the bee's egg which floats in a cell full of honey. The
larva is not yet at a stage at which it can enjoy the
honey—in fact, if it fell in the honey it would probably drown. It maintains, therefore, a precarious
position on the shell of the bee's egg, using it as a
kind of raft while it carries out its next transformation. It changes its skin, and with it, its diet. Now
it can eat honey and it lives comfortably for the
rest of its larval life on the food provided by the bee
for her own young.

At last the larva pupates and later still the perfect
beetle emerges. And now there is a gap in its
history. How does this portly, unathletic, slow-
moving beetle manage to waddle from the bee's
nest to the fields or commons on which it is generally found? Nobody knows, nobody has seen such a
thing happening. But happen it does, and so the
cycle starts again: the prodigious fecundity, the
prodigal wastage, the few, the very few, lucky
survivors. It has been calculated that out of ten
thousand eggs laid, perhaps two beetles will grow
to maturity. Nature is not an economical steward.

In fact, when it comes to slaughter, no living creature can compete with her at all, not even the Golden Gardener beetle. Ten thousand eggs laid, nine thousand nine hundred and ninety-eight lives cut off before maturity is reached! And this is happening every season to the offspring of every *Meloë* mother. Moderation is a quality seldom to be observed in the insect world.

4

THE UNDERTAKERS

DR. MUFFET remarks of the beetle that 'it is bred of putrid things and of dung and it chiefly feeds and delights in that.' Dung and corpses—they seem a singularly unattractive choice. But after all, we feed on corpses ourselves, even though we disguise them by cooking them first, and the Burying Beetles are undeniably useful in consigning putrid bodies decently underground.

The Burying Beetle is a handsome creature. He is black and wears two stripes on his back of a vivid vermilion or orange. There seems to be a difference of opinion about his smell for Fabre says he carries an odour of musk, while other naturalists maintain that his scent is so repellent and strong that clothes on which he has crawled smell fetid for days afterwards.

These beetles are attracted to a corpse by the smell as soon as it begins to putrefy. When they come on a dead bird or mouse, or whatever the find may be, the male expresses great satisfaction by wheeling round above it in the air. The female, of a more practical turn of mind, settles down to eat without any preliminary flourishes. The male soon joins her and they consume what must be a distinctly savoury meal until their hunger is satisfied.

The interment comes next. The beetles reconnoitre the surface of the ground and if it is suitably soft—a ploughed field, for instance—the burial can begin at once. This is chiefly carried out by the male. The female, again giving evidence of a practical nature, hides herself in the body and allows herself to be buried with it.

The excavations are made directly beneath the body. First of all, the beetle digs a furrow all round it, throwing the earth outwards, then he makes another and another, travelling inwards all the time. Soon the beetle disappears from sight underneath the corpse and the work can only be followed by observing the way it heaves up and down from time to time, and the steadily growing wall of earth which forms around it. As the wall gets higher, the corpse slowly begins to sink downwards into the hole beneath it.

At the end of about three hours or so, the beetle emerges, crawls on to the corpse and either rests or contemplates his labours for perhaps an hour. Then he begins again. The excavations continue and from time to time the beetle advances matters by pulling the body energetically downwards. After another three hours or so, the beetle again crawls up on to the corpse, looks round at the scene, calculates, perhaps, the work still to be done and then drops down as though suddenly struck dead.

It is only, however, the exhaustion of hard work for he soon rouses himself again and now he brings his task to completion. He settles the corpse into its little grave, treads it down firmly and then begins to push back the loose soil with his head. This takes only a short time and he then treads down the earth on top. He has now buried the corpse, and with it his own bride. It only remains to bury himself as well. This is the matter of a moment—he soon tunnels his way down and bears his mate company while she delivers herself of her eggs. This labour completed, they both indulge in a further and well-deserved meal before emerging and flying away on the scent of further carrion.

Theirs seems a strenuous and particularly blameless life. It is difficult, in fact, not to sympathise with Fabre's Burying Beetles, for there is no doubt that he teased them outrageously in his search for knowledge. He was sceptical, it seems, of the reputation they had won for intelligence. It was said that if a dead mouse lay on ground too hard for a grave to be dug in it, the perspicacious insects would remove it to another place where the soil was softer. So it had been said; but Fabre doubted it.

He decided to face his own Burying Beetles with this problem. He placed a brick in their enclosure and lightly covered it over with sand. So much was simple; but now he needed a dead mouse. This sounds as though it should have been simple, too, but, as he remarks, 'the moment a very common thing is needed, it becomes rare.'

At last, however, he secured the 'mouse of his dreams' and placed her on the centre of the brick. His seven Burying Beetles were resting under the sand but the enticing odour of carrion soon penetrated their dreams. Three of them—two males and one female—came hurrying up and disappeared under the mouse. The little corpse heaved and shook as the sand was cleared away from beneath her, and the problem—the brick—was revealed to the frustrated beetles.

They persevered for a little, but at last appeared to realise that circumstances were against them— that they were, in fact, so far as burial in that site was concerned, insurmountable. A male appeared from under the corpse, scratched a little on the surface, then disappeared under the body which began to sway about. 'Is he advising his colleagues of what he has discovered?' asks Fabre. 'Is he arranging the work with a view to their establishing themselves elsewhere, on propitious soil?'

Apparently not, so far as could be judged from their resultant activities. The beetles pushed at the mouse, with a will but not, unfortunately, in the same direction at once. The body swayed and oscillated violently, but in the end it remained within the little rampart of sand which the beetles had built up round it.

A male ran out again apparently to reconnoitre. He sank a trial well some distance from the body and returned to his colleagues. Activity was resumed, but seemed no better directed than before.

The other male beetle emerged to reconnoitre in his turn. Trial borings were made here, there and everywhere over the whole surface of the enclosure. Plenty of suitable sites for a burial had now been found, but no concerted effort was made to move the mouse towards any of them.

At last, by chance it seemed, the wall of sand was crossed and now there appeared to be some effort at co-operation. The body was carried at a fair pace to one of the regions which had been sounded and the burial proceeded without any further hitch. But it had taken the beetles just on six hours to decide that the mouse could not be buried where she lay and to remove her elsewhere. Fabre appears to have little patience with their stupidity, but surely he was demanding a good deal from them. A dead mouse, after all, is a not inconsiderable load for three beetles and the problem demanded a certain amount of resource.

Fabre soon demanded a good deal more of his beetles. He took from the kitchen an iron trivet, arranged a hammock of raffia between the three legs and placed the corpse on the hammock which was level with the soil. The beetles made little of this trial. Without difficulty they cut through the raffia on which the corpse became suspended and carried out the burial in good time.

Fabre, however, had not nearly finished with them. He suspended corpses on improvised spits, he entangled them on plants, he fastened them to upright twigs, he tied them up by the legs. His beetles tackled these varied tasks with unflagging enthusiasm but only limited credit. They were quite prepared to gnaw through the suspended leg of a mouse in order to release the body, but when a twig driven into the ground at a slant an inch or two away from the body was the real cause of the trouble, they appeared to possess no reasoning powers which would enable them to direct their assault against the twig, rather than against the obstinate corpse. In short, he found that if the problems which faced them were similar to those which they were liable to encounter in nature, they were able to solve them. If not, they were defeated.

Fabre concludes by denying them all intelligence and maintaining that they are guided by nothing but blind instinct. In this, perhaps, he is a little unfair. The intelligence of beetles may be so rudimentary as to be virtually negligible, but they do show at any rate *some* capacity to adapt their behaviour to changing circumstances—they do not act as though they were driven onwards by clockwork engines, blindly disregardful of any obstacles they may meet. If they were only able to solve such problems as they habitually meet under natural conditions, at least they have learnt, in the course of time, to solve these. It was demanding perhaps rather much to expect them to find a way of circumventing a brand-new sort of difficulty at the very first attempt.

Whatever may be said, however, of their intelligence or lack of it, their standards of behaviour showed a lamentable decline as the season advanced. Their domestic life had seemed wholly admirable in the spring when the cares of founding a family were occupying their energies. As the days grew hotter, however, the beetles lost interest in work and buried themselves below the surface of the soil. From time to time Fabre noticed that one would push his way rather feebly to the surface and always he would have some disability—a leg bitten off, it might be, sometimes two, and several of the joints missing on others. At last one wretched cripple was seen to drag itself miserably along with only one leg left out of the six. Another invalid emerged, whose plight was a little less desperate, and immediately proceeded to slaughter the first beetle.

And so their final days were passed. The consolation of work, the absorbing tasks of providing for the coming generation were finished and over. They exhausted what little remained of their dying vigour in attacking and devouring their companions.

5

A MIGHTY COMPANY

D R. MUFFET, who has always a kind word for almost any kind of insect, remarks of beetles that they 'serve for divers uses for they both profit our mindes and they cure some infirmities of our bodies. For though its house be but a dunghill yet it lives contented therewith and is busied and delighted in it. For it lives by the law of Nature and will not exceed her orders.'

The trouble with any generalisation about beetles is that there are so many of them. They form the largest division in the whole of the animal world and there are more than a quarter of a million different kinds of beetles. There are the huge tropical beetles, a good six inches in length, some of them equipped with menacing horns, and there are beetles so tiny that they look like specks on a piece of paper and can scarcely be discerned at all without a microscope.

Some of these tiny beetles settle down to live a social life in ants' nests. The ants appear to regard them as pets and behave as though they were really attached to them, feeding them and caressing them and rescuing them when danger threatens. Other beetles develop a relationship of mutual benefit with mealy-bugs. A pair of beetles, male and female, bore into the hollow leaf-petioles of a certain kind of tree when it is not fully-grown and begin to feed on a nourishing tissue which they find there. They are soon joined by small mealy-bugs which enter through the hole made by the beetles and settle in the grooves which have been produced by the beetles' energetic eating. The mealy-bugs apply their mouth-parts to the exposed tissues of the tree and suck its juices. Soon the beetles begin to breed and their larvae run about and live on the same food as their parents. Their diet, however, has now enlarged, for both the parent beetles and the larvae have learnt to stroke the mealy-bugs with their antennae and feed on the tiny drops of honeydew which are produced when their backs are suitably tickled. The beetles are very greedy for this food and spend hours milking their snow-white cattle. Sometimes a number of beetles stand round the same mealy-bug, like pigs at a trough, and quarrel among themselves for priority. The stronger beetles butt away the others with their heads, but they are not left long in possession.

The colony grows and enlarges and pairs of young beetles soon go off to start new colonies. Eventually, however, as the tree grows more mature, the ants take possession of it and drive out the beetles, while retaining possession of the cattle of the vanquished.

One of the commonest beetles is the lion-hearted Devil's Coach-horse, which cocks up his tail in the

air and in this attitude of defiance faces any and
every adversary, no matter how formidable he may
be. It is generally described as an ugly beetle—
in fact, one naturalist uncompromisingly declares
it 'the very ugliest insect in England'—but this
seems a little hard. It has a certain grace and
nothing seems to daunt its spirit. One observer,
Mr. J. G. Wood, describes how he came across this
beetle at the foot of the old Clifton Baths stairs at
Margate. He threatened it with the point of his
stick, but the beetle rushed to the attack with open
jaws and 'fought most valiantly.' Mr. Wood continued to threaten it and the beetle slowly re-
treated the whole way up the stairs 'with its face
to the foe and its jaws wide open,' pausing as each
new stair was gained to launch a fresh attack at
the stick. Mr. Wood was so pleased with this display
of spirit that he carefully placed the creature in a
place of safety when he reached the top.

The larvae of this beetle are quite as fierce as
their parents and even more dangerous, for their
appetite is larger. They attack and kill any insect
which comes within their reach, even their own
kind. They seize hold of their prey by the soft
chink which can be found between the head and
neck, sink in their powerful jaws and suck the body
dry.

Another savage larva is that of the Tiger Beetle
which lurks at the bottom of a pit after the manner
of the larvae of the Ant-lion. It digs its pit and suspends itself at the entrance by means of a pair of

hooks situated about half-way down its back. Its head just blocks the entrance and it waits patiently until some incautious insect passes overhead. Then it seizes it in its jaws, drags it to the bottom of the pit and devours it at leisure.

Blaps is another beetle which is generally described as repulsive, this time with considerable justice. It is a large, sluggish creature, dull black and bloated, which crawls about in cellars and damp places and has a smell as repulsive as its appearance.

The Stag Beetle is a most ferocious-looking creature and the largest beetle seen in this country. Yet it is harmless. The huge, antler-like jaws of the males are reserved for wrestling-bouts among themselves. And even in these battles, no harm is done and the warriors emerge unscathed.

Other unpopular beetles batten on furniture and eat their way through books. The Death Watch Beetle has a fancy for old wood and has won a romantic reputation as a foreteller of imminent doom. Its solemn ticking, heard in the silence of the night, has long been thought to announce the death of one of the inmates of the house. The significance of this noise, however, is more truly romantic, for it is a mating call and is used as a means of signalling by the beetles as they seek for one another.

Some beetles live most of their lives in the water and, hugging little air bubbles to their chests, they plunge bravely down to seek their prey in ponds. The Great Diving Beetle is one of the largest and commonest of the water beetles and is a pitiless destroyer of any insect it can find. Its larva is equally savage and of a distinctly alarming appearance with its curved and pointed jaws, which look rather like tusks. Its practices are as unpleasant as one might guess. The jaws are tubular and when they have pierced the victim, the larva injects through them a special kind of fluid which has the sinister property of pre-digesting its destined meal, even though the meal is by no means dead. The tissues of the unfortunate prey are slowly dissolved and the larva sucks them up in the form of a nourishing soup.

Dr. Muffet, with all his affection for the insect world, seems to have been hard put to it to find something kind to say about the external attractions of beetles. 'Most of the beetles are hideously black,' he observes, 'yet some have their cases shining with a blacker, others with a more pleasant green. There are some also that shine like gold. Some there are which fly about with a little humming, some with a terrible and with a formidable noise.'

There are, in fact, all kinds of beetles, pleasant and unpleasant. The varieties are endless and it is impossible to come to any general conclusions about insects differing so widely in their habits and their ways of life. And yet, when all is said, one is left with a feeling of affection for beetles. One remembers the common black beetle of the garden, plodding laboriously through the soil, the shining beauty of the sweet-smelling musk beetle, the unassuming lady-bird, destroyer of garden pests. One remembers the curious glow-worm, switching on her light to draw her love to her side and discreetly extinguishing it again when he reaches her—though none knows how she does it; and the alarming Bombardier Beetle, defying his enemies with volleys of artillery accompanied by audible pops and puffs of smoke. Above all, in this remorseless savage world of the insects, one salutes the domestic virtues of that humble Dung Beetle, the *Sisyphus*, so attentive to his mate, so hard-working for his children.

THE SOLITARY WASP

1

SWEET *AMMOPHILA*

THE yellow and black striped wasps which infest
the tea-table in summer live in communities
in much the same way as the honey-bee. There
are other wasps, however, which follow a life of
independence and individual enterprise. Alone
they construct their nests, alone they stock them
with provisions for the coming generation, alone
they lay their eggs. It is true that they do not
conceive their eggs quite alone—the male is
allowed to play his necessary part in this enterprise—but he is allowed to do little else. For the
most part, these wasps seem to have no leanings
towards family life, or even towards a division
of the labour of founding a family. The female
wasp is a natural solitary.

She is a graceful creature and her colouring is
often attractive. Some hunting wasps have yellow
and black stripes, others are red and black, orange
and black or just black. Her methods of providing
for her young seem peculiarly macabre even when
judged by the savage standards of the insect world,
but what with her 'pretty ways' and her varied
and fascinating habits, she seems able to inspire

an unusually fervent admiration among her observers.

'In studying the species that come our way,'
write Mr. and Mrs. Peckham, the American entomologists, 'we are continually developing distinct
likings for some kinds above others and when the
season's work is over we remember them with
lively pleasure. It is thus, dear little *Oxybelus*,
that we dwell upon the thought of you and your
pretty ways. No other wasp rose so early in the
morning, no other was so quick and tidy about her
work, no other was more rapid and vigorous in
her pursuit of prey.'

They also admired *Pompilus*, that 'gay, excitable little wasp' and thought her 'delightful to
meet.' *Ammophila*, however, 'most graceful and
attractive of all the wasps,' was their favourite.
They were impressed by her intelligence, appreciated her 'distinctive individuality' and even ex-
pressed gratitude for her 'obliging tolerance' of
their company. The solitary wasp, in fact, emerges
from the glowing descriptions of the Peckhams as
a wholly admirable insect, almost without faults.

It is hard to remember, when they praise these
charming, attractive, beguiling creatures, that they
are speaking of the ogresses of the insect world.
It is, in any case, a world of butchery, of cold
cruelty, seldom lightened by any flash of what we
might call mercy or affection. Yet slaughter is
usually carried out to assuage hunger, not for
malice, and death, when it comes, is sudden, the

anguish of a moment. This is not so for the victims of the solitary wasp. For them, there is no quick escape from life.

As far as her own tastes go, the solitary wasp is a vegetarian. She sips nectar from the flowers and needs no nourishment beside. But her growing children need meat, and it is her concern to provide for them meat which will not putrefy, meat which will remain fresh for several weeks on end until their larval stage of development is safely completed. So what could be better for her helpless young ones than living meat, prey which is fully alive, but paralysed? This is not beyond her powers to provide, for she is armed with a sting which will do this very thing for her.

All the hunting wasps are concerned with catching insects on which their grubs will feed, storing this food and laying an egg on or beside it. With that variety of habits which adds to the fascination of wasp-watching, one species, *Pompilus*, keeps to this order of doing things, whereas another, *Cerceris*, excavates a burrow before catching prey, and *Eumenes* not only prepares its burrow, but lays an egg in it before adding the food. *Pompilus* excavates a new burrow for each large spider it catches, in contrast to *Cerceris* which stocks several insects in each of a series of branch passages at the bottom of each burrow. *Eumenes* is a collector of caterpillars, as a rule, and stores dozens of them in each series of cells, but one species has come to specialise on the insect which causes the cuckoo-spit on plants.

The solitary wasps seem to vary a good deal in the amount of energy and skill they bring to the task of digging a burrow—they are nothing if not individualists. *Ammophila*, however, does not aim very high. Her nest is hardly ambitious in design— it is a narrow tunnel in the sand about an inch long which widens out into a little pocket at the bottom. It is, as Fabre remarks, a 'poor lodging,' which is hollowed out at the cost of perhaps a day's labour.

The wasp digs with her mandibles and front legs. Some show a painstaking efficiency in this operation, but others seem so indifferent and casual in their methods that it can only be by a fortunate chance that the young confided to their nests ever survive. Soon the wasp has dug down so deep that she is hidden under the ground and she now reappears from time to time to carry away earth. To do this, she backs out of the entrance, flies a short distance and then gives a sort of flick which shoots the crumbs of earth she is carrying some distance away from her. Then she drops to the ground and appears to be taking a moment's rest before going back to her labours.

At last the nest is dug to her satisfaction and she goes off to collect a suitable lump of earth or small pebble. This she arranges with some care over the entrance. For most wasps, the temporary closing of the nest completes the business of nest-building, but others are perfectionists. The Peckhams describe one *Ammophila* who seemed never to be satisfied with her own efforts. First of all, she went off, found a largish lump of earth, brought it back and laid it over the entrance hole. It covered the hole quite satisfactorily, but the wasp appeared unhappy. She flew away, but she came back ten minutes later and settled by the entrance, seeming to inspect her nest with a mildly worried air. The entrance was all but invisible, but she none the less tore away the covering, carried out a few more loads of earth from inside the burrow and then began to hunt for a new covering for her nest.

After a time, she came back with a lump of earth, but seemed to lose confidence in its entire suitability as soon as she came in sight of the nest, for she dropped it and made off again. After some fairly prolonged hunting, she found a piece which was a perfect fit and placed it in the entrance hole. But still she could not rest. She brought another slightly smaller piece of earth and painstakingly arranged it above the other, but a little to one side. She stood back and gazed at her handiwork. 'It seemed to us,' record the admiring Peckhams, 'that we could read pride and satisfaction in her mien.'

They were wrong. The unhappy creature was inexorably driven by her urge to attain impossible perfection, and two hours later they found her still trying one pellet after another as a door for her nest—by that time there was quite a little pile of them lying at the entrance. The Peckhams say that she never succeeded in stocking her nest with a caterpillar, for they opened it several days later and it was empty. 'Perhaps she came to some untimely end,' they write. It seems more likely that the wasp was still searching unavailingly for the one perfect caterpillar.

Not all these wasps are so hard to please, however. Most of them close up their nests ingeniously enough with two or three lumps of earth or perhaps a stone, kick some dust over the top as an extra precaution, and then go off in search of prey without further worry.

Wasps are intensely conservative about the food they provide for their young—they provide the food that their mothers have provided before them for generations back and nothing else, apparently, will do. One species captures flies, others spiders, bugs, beetles, locusts, bees and so on. *Ammophila* appears to believe that her children will grow and thrive on either grey or green caterpillars.

Nobody knows what it is that makes wasps so constant in their feeding habits. The larva which feeds on the captured prey eventually spins a cocoon and passes through a state of sleep or trance before emerging as a fully-grown wasp. As a wasp, she has no taste for the meat on which she has lately fed. And yet, when she must provide for her young, it is meat of this kind and none other which she hunts and captures. Is she guided by some inherited instinct which somehow persists from one generation to the next? Or does she carry with her a dim, formless recollection of the taste and smell and sight of the victims which stocked her own nursery? Nobody knows; but it should be possible to find out. Dr. Bristowe has suggested that if a larva could be found at the moment of hatching or soon afterwards, and induced to eat some different food from that which its mother had provided, it might grow up to be a wasp which would hunt different prey from the rest of its tribe. If this happened, it would show that it is indeed some faint remembrance of larval days which guides the wasp in its choice of food for its young, rather than inherited instinct. It would also open up the possibility of training these wasps to attack garden pests. However, the experiment, though possible, would be extremely delicate and difficult to arrange, and it may never be carried out.

In the meantime, *Ammophila* remains fixed in her preference for caterpillars. One of the particular caterpillars she fancies lives underground, but she is expert in finding it out. Fabre and his whole family hunted for them without success for hours one day until he at last decided to follow the lead of a hunting *Ammophila*, and she guided him to the lair of one caterpillar after another. Her difficulty appeared to lie not in locating them, but in digging them out.

Ammophila scurries over the soil, when she is out hunting, lashing the ground with her antennae, occasionally pausing to scratch it a little. Her prey is difficult to come by and the Peckhams estimate that she can hardly average more than one caterpillar a day. But she is indefatigable. She continues her search hour after hour, concentrated and alert. Some observers have been lucky enough to see a catch and have described what happens.

Ammophila first begins to scratch at the bottom of a plant, and to pull up little blades of grass by the roots. Next she pokes her head into the ground, running excitedly here and there rather like a dog on the trail of a rabbit underground. And this is where the caterpillar makes a serious mistake. Whether it is interested by the noise going on overhead and wishes to investigate what it is, or whether it is panic-stricken at the nearness of the wasp and hopes to escape, no one can say; it makes little difference in the event. The unwary creature appears above the ground.

Instantly the wasp seizes it by the back of its neck. The caterpillar, a large and powerful creature, frantically coils and uncoils itself in desperate, frenzied contortions. The wasp is flung off. She attacks again, and again she loses her hold. The caterpillar, appearing to realise its mistake in ever coming out at all, makes frantically for the nearest cover; but the wasp is too quick. This time she gets a secure grasp on it and nothing will shake her off. Her mandibles are firmly sunk in the back of its neck, and she waits calmly, with an air almost of detachment, until the desperately writhing coils of her victim give her the chance she is waiting for. She curves the end of her abdomen under its body and the sting goes home.

Such encounters have been described by many naturalists, and the sequence of events after the sting is once administered shows little variation. The struggles of the caterpillar die away. It lies helplessly beneath the wasp, limp and inert. And at this moment, the victorious wasp leaves her captive.

She now hurls herself into a mad, disordered dance—so wild are her movements that Fabre, on first seeing it, thought that she must have been poisoned in some way and was in her death throes. She stamps on the ground, she flings herself down, frantically twitching her limbs and fluttering her wings. She gets up, and again she throws herself down with furious abandon. She lays her forehead on the ground, she rears herself up on her hind legs. It is her victory dance, a wild, ecstatic burst of triumph.

It is soon over. The wasp smooths her wings and curls her antennae. She gives her face a conscientious washing with her front legs. And then, her toilet completed, she is once more the sober matron preoccupied with duties which demand her urgent attention.

First of all, she gives the caterpillar several more stings, working carefully down the segments. It is advisable for her children's safety that the power-ful creature should be satisfactorily paralysed beyond hope of revival. She then takes the caterpillar's head in her jaws and crunches it slightly. It is an operation to which she appears to give the most careful and concentrated attention. This is no more than one might expect, for great delicacy of touch is required. It is her aim to induce a temporary insensibility in the caterpillar which will deprive it of the use of its mandibles without depriving it of life. A lusty caterpillar, capable of snapping at its captor or seizing hold of blades of grass and the like as it is dragged along, would be an embarrassment on the journey home. On the other hand, a lifeless caterpillar will certainly be no use as fresh meat for children not yet hatched or even laid. A delicate balance must be preserved between the two extremes and this the wasp generally achieves.

Next she gets a grip on the bulky caterpillar and begins to drag it laboriously home. In order that this phase of her operations may be completed successfully, it is naturally important that she should remember where home is to be found, and entomologists have interested themselves a great deal in the problem of how the wasp finds her way.

Fabre captured twelve female wasps and put each one separately in a paper bag. With this curious cargo, he walked about a mile and a half from the nests, marked each of his captives with a spot of indelible white paint and then released them. They flew only a few yards at first, settled on blades of grass and 'passed their fore-tarsi over their eyes for a moment, as though dazzled by the bright sunshine to which they had suddenly been restored.' Almost immediately after that, however, they took flight all in one direction, straight towards their home. A later inspection showed Fabre four of his twelve marked wasps busily at work by their nests, and he thought it fair to assume that the others were probably out hunting or underground in their burrow, rather than lost.

Next he put nine more wasps in paper bags, enclosed the lot in a dark box and set out for the nearest town, about two miles away. This batch was released in the middle of the street, in conditions which must have been confusing for the country-bred insects. However, each wasp in turn rose straight into the air above the houses, then made off to the south, the right direction for their nests. The majority of these, too, he was able to find on his next inspection of their nests, and the others may well have been occupied on their normal business—it would be unlikely that he would find all of them at work by their nests at the same time.

Experiments, then, have shown that these wasps are able to find their way over fairly long distances, whether by memory or by instinct. The problem of recognising their nests when they have once arrived fairly near them is another matter, however, for they are almost invisible. Some experiments suggest that the wasp remembers landmarks near the nest—a tuft of grass, perhaps, or a bush. Some pieces of heather stuck in the ground by one wasp's nest were effective in preventing the owner from being able to find it. As soon as the heather was removed, however, she found it, apparently without any difficulty at all.

Sight certainly seems to play a part in guiding the wasp as soon as she is anywhere near her nest. One wasp dragging her prey home had her view obscured by a large bush of broom which lay directly in her path. When she got up to it, she laid down her prey and flew up in the air so as to have a look over the top. She then descended and began to drag the victim through the bush, but she none the less felt obliged several times to fly up in the air, apparently to make sure she was still on the right path.

Some observers seem to think that colour has quite a lot to do with recognition. Adlerz saw one wasp, which had dug its nest near a red plant, examine several other red plants before finding the right one. When once wasps are really near the nest, however, scent probably helps them.

Wasps show some powers of memory in dealing with their prey, too, for many of them habitually hang it up on a twig or something of the kind, out of the reach of ants, while they go off to inspect their nest before returning to their prey and dragging it inside. They are evidently able to remember where they have left the prey and the path by which they must return to it. Some can even carry a fairly accurate recollection of size in their minds. The Peckhams describe a wasp who found that her prey—a spider—was too large to fit into the entrance of her nest. She hung it up among some clover blossoms and then 'washed and brushed herself neatly.' Apparently she was in no hurry to complete her business, for she took 'several little walks' and altogether dawdled away fifteen minutes.

At the end of this time, however, she had not forgotten what she had to do. Nor had she forgotten her estimate of the spider's size. She went to her nest, enlarged the entrance as much as was necessary, but no more, and then went to fetch the spider. Her estimate of the amount of enlargement needed proved to be exactly right, for she was now able to drag the spider inside, but with no room to spare.

Individual wasps seem to vary in their powers of memory and observation, but the majority appear able to find their nests with no particular difficulty, whether they do it by sight, scent or some other sense. The caterpillar, as it is dragged along by its captor, can have little ground for hoping that it will never arrive, even if it were capable of such thoughts. *Ammophila* will surely find her nest and pull the caterpillar inside after her. And when this happens, the caterpillar can by no means escape the nightmare which lies before it. The wasp lays her egg upon the caterpillar, goes out of the nest, and shuts the door behind her.

Ammophila, in fact, generally shuts it with painstaking efficiency. The Peckhams relate the story of one *Ammophila* whom they remember, they say, as the 'most fastidious and perfect of the whole season, so nice was she in her adaptation of means to ends, so busy and contented in her labour of love, and so pretty in her pride over the completed work.'

She filled up her nest by putting her head into the opening and biting away the loose earth at the sides so that it fell down to the bottom. She let a certain amount fall down in this way, then she jammed it firm with her head and did the same thing over again. When the nest had been filled up level with the ground, she went and selected a small pebble. Holding this in her mandibles, she used it as a hammer to pound down the earth until it was as firm as the soil surrounding it. 'We are claiming a great deal for *Ammophila* when we say that she improvised a tool and made intelligent use of it,' write the Peckhams, 'but fortunately our observation does not stand alone.'

The dear, intelligent little creature flies at last upon her way, and in her nest, so splendidly secured, events unfold themselves. The caterpillar lies in its narrow surroundings, paralysed, helpless, but fully alive. In a few days, a larva emerges from the egg which has been planted on its chest. The larva is hungry—it will appreciate the meat which its attentive mother has provided. It does appreciate it. It bites into the caterpillar on which it is lying.

The Peckhams have observed this to happen, and they say that the caterpillar 'reared up on end.' This is disquieting. It would be pleasanter to

believe that the caterpillar is anaesthetised as well
as partially paralysed. But does it in fact feel pain
as the larva progressively devours its living tissues?
Nobody knows. Who can estimate the degree of
pain in the consciousness of an insect? How is it
possible to interrogate it on this point? And whether
it feels pain or not, there is nothing it can do. The
paralysis produced by the sting of the wasp appears
to vary a good deal in severity, but the caterpillar
certainly cannot move freely enough to shake off
the persistent little larva. In any case, the larva has
soon plunged itself right inside the caterpillar's
body in pursuit of its meal.

This meal continues a long time—two weeks,
three weeks. And the caterpillar lives a long time,
too. It lives, in fact, to the very end, for the larva
is guided by a wise discretion in its eating. By the
light of instinct, by the processes of evolution, it is
prompted to eat the less important parts of its living
meal first and to leave the vital organs to the last,
so that its victim shall not die too soon and go bad.

In time, however, the meal is finished, the caterpillar is released from its long ordeal, and the larva
changes itself into a pupa amid the drained and
empty shell which is all that now remains of its banquet. The next stage is passed without food, and at
last, after a suitable interval, the fully-developed
and now vegetarian wasp emerges into the sunshine.

2

LONG-SUFFERING *BEMBEX*

Some kinds of wasp are not able to finish with their nests when they have deposited their eggs, for they are obliged to re-enter them at frequent intervals. *Bembex*, for instance, digs her little cell in the sand and stocks it with a single fly on which she lays her egg. There is no need for her to hammer down the entrance with a stone and make it firm—it would be highly inconvenient for her if she did so. For the single fly will by no means suffice to nourish her growing offspring.

Instead of closing up the entrance effectively, therefore, she merely covers it over with loose sand which will give way easily when she wants to thrust her way back into the cell. For a few days after laying her egg, she can relax. She may pass the time by sifting the sand in front of the door, and removing small pieces of gravel or pebbles which might get in her way if she wanted to enter in a hurry. For the most part, however, she drinks nectar from the flowers or suns herself happily on the scorching sand. She is not without company, for the *Bembex* wasp lives in a kind of village—a colony of nests built only a few inches apart. This

106

propinquity does not appear to inspire affection, however, for, according to the Peckhams, these wasps are always quarrelling and attacking each other and are 'certainly very unneighbourly.'

A day or two passes and then, prompted, writes Fabre, by the 'instinct of a mother,' the *Bembex* wasp appears to realise that her young one needs more food and she goes off in search of a fly. She returns with her prey, penetrates her nest, deposits it near her larva and goes out again to wait. The waiting interval this time is much shorter, and it becomes shorter every day. The 'instinct of a mother' is apparently equal to judging the regular increase in the appetite of a rapidly growing larva, and every day the wasp brings more and more flies to the nest.

For nearly a fortnight *Bembex* is thus occupied in bringing provisions for her young, and towards the end of that time she is very busily occupied indeed. The Peckhams succeeded in rearing a *Bembex* larva and they record that it ate forty-three flies in five days. Fabre gave one eighty-two flies in eight days. The gluttonous larva, so lavishly fed, became fat and bloated, and crawled 'heavily along with his great lumbering belly among the scorned leavings,' says Fabre, 'rejected wings and legs and horny abdominal segments.' At last it pupates and the mother sets herself to repeat the business all over again.

Bembex is one of the few wasps to tend her young after they have been hatched. This is to be regarded as an advance—the faint beginnings of family life, the first frail blossoming of maternal care. *Bembex* has other cares, too. She is plagued and terrorised by a small, nondescript-looking fly, a feeble and insignificant creature called the *Tachina*.

The grounds of her terror seem a little mysterious. The *Tachina* is a fly, and *Bembex* preys on flies. Why then does she not catch this fly, as she catches others, and destroy it? She can fly fast enough to overtake it, and she is certainly powerful enough to murder the little creature with the greatest of ease. But she does none of these things. On the contrary, she runs away from it in panic.

Although it is surprising that the wasp should be afraid, it is not surprising that she should feel hostility towards these small, malevolent creatures, for it is their ambition to lay their eggs upon the prey which the wasp is carrying to her nest for her own young. They are anxious to provide for their families, too, but at another's expense. This demands from them skill, concentration and daring.

They have all these gifts. And they have patience as well, for they lie in wait on the sand hour after hour until the wasp shall return at last to her nest and they can achieve their purpose. They stay quite still with their eyes fixed unwinkingly on the wasp's burrow. The hours pass, but they never move, they never relax. They are ungrudging of their time.

The wasp arrives carrying her prey, and at once there is a slight, betraying movement among the flies. Their great eyes, the colour of dried blood, are fixed upon the wasp, their heads slowly turn as they follow her with their gaze.

At once, she is disturbed. She has seen them. She hovers doubtfully over her nest, and the anxious beat of her wings gives out a plaintive, whimpering note. She hesitates for a moment, and then drops straight down and hovers a few inches above the ground. The flies wheel up in flight and take their positions in line behind the wasp.

She turns on them; but they have turned with her. With military precision, they have kept in line. The wasp turns again, she advances, she retreats, she flies up, she flies down, she flies slowly, she flies fast. The flies are not put out. They keep in line behind her, whatever she may do.

The wasp, defeated, alights on the sand. Her followers settle behind her, motionless, alert. Suddenly, she darts off again with a shrill note of anger and despair. She flies far away from the nest. The flies let her go, for they know she will come back. They settle again to wait, their blank, expressionless eyes fixed on the burrow, their dull grey bodies inconspicuous against the sand.

They are the assassins of her children, the destroyers of her family, and she lets them live. Sometimes, indeed, she succeeds in chasing them away for the moment, but this gives only a temporary respite. They come back. It seems so foolish, so reckless. *Why* does she not kill them? The fact is, the wasp is a creature of habit, hampered and enchained by instinct. The explanation may well be that her instinct prompts her to use her weapons of attack only for the slaughter of prey, and these flies do not provide the necessary stimulus. Besides, when she encounters them, she has generally secured her prey and the instinct to kill is, for the moment, finished and gone. Fabre, however, has a simpler explanation than this. He says, 'Since this wretched little fly has her tiny part to play in the general order, the *Bembex* must needs respect her.'

However that may be, the wasp in time returns to her nest. The manœuvres repeat themselves all over again, until at last the mother's patience is worn out and she enters her nest, dragging her prey after her. This is the moment—the split second— for which the flies have been waiting; the moment when the wasp is half in, half out of her nest and her movements are hampered; the moment when her disappearing prey still projects from the burrow.

The flies, displaying an impressive control over their natural functions, sweep down upon the wasp's prey and manage to lay their eggs upon it without a second's delay. This tricky task accomplished, they settle down on the sand once more 'to meditate,' as Fabre puts it, 'fresh deeds of darkness.'

Has the wasp any realisation of what has happened? She behaves as though she had not. She continues her routine apparently unmoved, though she must now provide not for one hungry mouth, but for perhaps a dozen. And if she is concerned for her larva's welfare, she will see to it that the provisions are plentiful. For as long as there is enough for all, relations between host and parasites may continue amicable; but for no longer.

The parasites' maggots develop more quickly than the wasp's larva. They take more than their share of the common food, and grow strong. The larva, however, appears to take all this in good part and shows no resentment. Nor do the alien grubs show any enmity on their side. Danger, however, is not far away.

The unfortunate wasp is simply not equal to the output now demanded of her. She had as much as she could do to provide for her own child. Now there are twelve others as well, all hungry and growing fast. And even if she were capable of feeding such a family, it is improbable that she has any realisation that an extra effort is needed. There is little doubt that she provides the usual rations and no more. And in time the inevitable happens. Food which would have nourished one admirably, is inadequate for this large brood. The hardy parasites manage on what they can get and in time they pupate. It is the true child of the house which goes hungry and suffers want. Weak and sickly, an undersized, underfed little wisp, it tries at last to spin its cocoon, but it has no silk. It dies among the healthy pupae of the parasitic flies.

This is not all it may suffer. If the provisions brought by the mother should not be plentiful enough for these uninvited guests, if she should fail to bring them on time, they attack the wasp larva and eat it up. They are untroubled by guilt, they feel no pity—the larva is food, and they are hungry.

Since the mother constantly revisits the cell bringing fresh provisions, it might be expected that, however limited her powers of mathematics, she would notice that her family had unexpectedly jumped from one to twelve, and that the greedy strangers, jostling for the best share of the food, are not in the least like her own children. Why does she not attack them, destroy them, hurl them from the nest? It is clear that no such thoughts even pass through her mind—if she may be said to possess a mind. She tolerates the strangers and she feeds them, just as though they were her own. Perhaps she even prefers them. For it is true that her own larva has become a weak and weedy creature, certainly no credit to any mother. And whatever she may feel about her unsatisfactory offspring, it is quite certain that all that will ever emerge from that nest alive are the progeny of the insignificant, treacherous flies.

3

INELUCTABLE INSTINCT

The wasp, although she may occasionally modify her behaviour to suit changing circumstances, seems for the most part to be guided by a series of uncontrollable urges which succeed each other with a strange, mechanical inevitability. This has been demonstrated by a number of experiments. One of them was carried out by Fabre and his subject was the *Bembex* wasp. He uncovered a *Bembex* nest all the way along so that it appeared like a sloping, open trench with the little cell exposed at the end of it. The door, of course, had gone with the rest, but the larva remained in place among its heap of dead and dismembered flies. Fabre awaited the arrival of the mother.

At last the wasp arrived and went straight to her entrance door—or rather, to the place where the entrance door should have been. She seemed disconcerted by the new appearance of her nest and spent more than an hour digging and sweeping the sand away in search of the lost door. She seemed confident of her judgment of locality, for she confined her search to a few inches of ground and swept and hunted over and over again in this small area.

113

She never showed any disposition to explore the
trench, although her frenzied hunt for the door
took her along it once or twice. Her attention was
never attracted by the wriggles of the larva which
was writhing in discomfort from being exposed to
the fierce heat of the sun. Always this mother
returned to the spot where she *knew* her door must
be, for at this moment, as Fabre remarks, she wanted
her entrance door, the usual door and nothing but
that door.

In her wanderings, the mother at last came upon
the larva. This, one might have thought, would be
the end of her search. She only required her door
in order that she might reach her larva, and here
was her larva before her eyes. 'At this moment of
meeting after long suffering, have we a display of
eager solicitude, exuberant affection, any signs
whatever of maternal joy?' asks Fabre. The answer
is no. Far from it.

The mother failed to recognise her larva at all.
She walked over it absent-mindedly, and trod it
down as she hurried about. Once, when she wanted
to dig in the cell, she pushed it aside with a 'brutal
kick.' A more painful scene still was to follow.
The larva was at last roused to retaliation and
seized its mother's leg in its powerful jaws. There
was a short, bitter struggle. Then the larva let go,
and the alarmed mother flew away, 'making a
shrill, whimpering noise with her wings.'

The battle between mother and son is unusual,
remarks Fabre, but the mother's indifference is

invariable. She always continues her passionate search for the missing door, apparently quite unmoved by the sufferings of her offspring, which is thus doomed to die without succour, although its mother is at hand.

Such are the limitations of instinct. The wasp is quite unable to miss out one link in the accustomed chain of events. First she must have the door, then the passage, then the larva. To put it more scientifically, the wasp's normal behaviour gives the impression of 'a chain of reflexes, each one of which releases the next.'

Pompilus, another kind of hunting wasp, seems no more able to cope with the unexpected than *Bembex*, although she is greatly admired by naturalists. 'We look back with much pleasure on our acquaintance with this gay, excitable little wasp,' write the Peckhams with their usual enthusiasm. 'She was so full of breezy energy that it was always delightful to meet her.'

So long as her routine is not disturbed, *Pompilus* seems to carry out her business with brisk efficiency. Spiders are her prey and she looks for them among low-growing vegetation, on the stems of plants or on walls and tree trunks. One kind hunts the wolf-spiders which dig little burrows in the sand and close the entrances with silk. Dr. Bristowe has described the behaviour of a *Pompilus plumbeus* hunting for one of these spiders. She scurried quickly across the sand, but suddenly stopped and began to scratch with her jaws and front legs. 'She now behaved,' he writes, 'just like an excited dog in search of a rabbit.' She scratched and circled and scratched again, vibrating her antennae and touching the ground with them, so as to check the spider's whereabouts. She had no luck, however, for these spiders are often able to escape from their burrows by a back entrance while the wasp is besieging the front door, and that is what happened.

On fine days, however, the entrance to the spider's burrow may be left open and then the wasp dashes straight in. This seems fearless conduct, for the spider is a powerful creature and should have the advantage on its home ground. Such is their inherited fear of wasps, however, that if once they are cornered, they make no attempt to defend themselves at all, but appear to resign themselves without struggle to their very horrible fate.

The wasp emerges with her captive, carries it aloft in front of her for a short distance, excavates a shallow pit and buries the spider. Next she rushes off and begins to dig her nest close by, rapidly kicking out the sand with her back legs.

Pompilus is a fast worker and may complete her nest in as little as five minutes, although some of these wasps behave as though they suffered from doubts and begin, and then abandon, one nest after another. When it is finished, she goes straight back to the spot where the buried spider lies, digs it up and drags it near to the entrance of the nest. She then drops it again and inspects the burrow. When she is satisfied that it is in order, she pulls the spider to the entrance and then goes in herself backwards, pulling the spider after her. She has sometimes been observed to do some more excavating after she has got the spider inside, and some observers think that she does not complete the nest until this stage is reached.

Eventually, however, she emerges, after first laying her egg on the spider, and she fills up the burrow with neat efficiency, even taking considerable pains, as it seems, to remove all signs of digging and disguise the site of the nest. All this work seems to be well-organised and intelligently carried out. It suggests a superior insect which might well be able to meet untoward circumstances with resource. Yet she seems to be just as much the slave of instinct as *Bembex*.

The English entomologists Richards and Hamm relate that they removed a spider from the nest of a *Pompilus* wasp with forceps before the nest had been closed, and placed it at the entrance. The wasp returned, touched the spider with her antennae, and then closed up the empty nest just as though the prey had been safely stored inside.

The Peckhams disturbed a *Pompilus* wasp at an earlier stage of the proceedings by removing her paralysed prey which had been hung up on a bean plant, and substituting an undamaged spider. The spider, alarmed at being handled, kept perfectly still, and, when the wasp returned, was as motionless as the paralysed spider it was replacing. The wasp, however, was not deceived for a moment.

She took no interest in the substitute spider, but hunted fruitlessly for her own for some time. At length, she gave up the hunt and flew away in search of fresh game.

It may have been clever of her to see through the trick which had been played on her, but surely it would have been sensible to accept the spider that offered itself rather than go off in search of a problematical new one? It may be, however, that the appearance of the frightened spider hanging on the plant was different from that of the spiders normally hunted, and so the right instincts were not called into action which would have made her sting it. Even a ready-paralysed spider was no more tempting. The Peckhams found an ant in the act of stealing a paralysed spider which had been left unguarded by its owner. They took it and placed it on the doorstep of a wasp which was away from her burrow hunting. The wasp returned with a victim of her own and seemed to regard the Peckhams' offering as no more than an obstruction. She did a little digging beside it and finally seized it by the leg and dragged it out of the way. Then she stored her own spider and flew away, apparently in search of fresh victims. There is no reason why she should not prefer the prize of her own skill to any other spider. It is natural enough. But even so, it would surely have been prudent to use the other spider, which was perfectly fresh and quite at her disposal, to stock her next nest and so save time and energy.

Other kinds of hunting wasp seem no more

adaptable. Fabre tried removing prey from the burrow of a *Sphex* wasp while she was engaged in closing it up. The wasp, who had been disturbed at her work, returned and found that the door which had been partially closed was now open. She went inside and stayed there for a few minutes. Then she emerged and proceeded with her work of closing up the burrow, exactly as though she had never been interrupted. When this was done, the insect brushed itself, says Fabre, seemed to give a glance of satisfaction at the task accomplished, and finally flew away. And yet she had been inside the burrow, she had had the opportunity of noticing that the prey, and with it her egg, had been removed. Why did she bother to close up this rifled nest with conscientious care? Presumably because she had reached this stage in the cycle of instinctive behaviour and she had no powers of reasoning to persuade her that her efforts were now futile.

Dr. Bristowe records the same sort of behaviour in two wasps which had dropped their prey accidentally as they entered their burrows. Unaware of what had happened, or unmindful of its significance, they proceeded to stay inside their burrows for the normal length of time required for storing a victim. He also quotes a third example. This was provided by a *Cerceris* wasp, which saves itself the trouble of digging a burrow from the beginning by taking possession of a bee burrow and enlarging it.

This particular kind of bee is the natural prey of the *Cerceris*, but the owner of the stolen burrow is

never used for this purpose by the wasp, since, at
the time when they meet, the wasp's instincts are
entirely concentrated on digging, and the hunting
instinct is dormant. Dr. Bristowe watched one of
these bees emerge after an encounter with a wasp
below ground. It had evidently been stung and
could neither fly nor walk properly. It fell down
in fact, and rolled back into the burrow from which
it had crawled. The next morning it was lying
outside the burrow again, making feeble movements with its legs. It seemed likely that the wasp
had thrown it there. And yet this was exactly the
prey which the wasp would go out to hunt as soon
as the burrow was complete. The explanation
seems to be that the wasp must do all things in due
order. First she must dig her burrow, and only
then can she interest herself in the capture of prey.
And even in the hunt there is no elasticity—prey
must be captured in the accustomed way above the
ground. An inert body lying ready in the nest is
powerless to call forth the appropriate instincts.

The wasp, it seems, is the slave of her time-table.
It is impossible for her to jump a stage and so save
herself trouble. It seems equally impossible for her
to go back a stage and carry her prey down into
the nest again when it has been removed and
placed at the entrance. The fingers of her internal
clock have moved on and the time for such activities
has gone by.

4

SEX AND DOMESTIC LIFE

THE males play no part at all in the strenuous work of hunting and provisioning which occupies the female wasps. Their life is spent in frivolity and pleasure—feasting, dancing in the sunshine, love. 'Male wasps,' remarks Bristowe somewhat severely, 'hover round the burrows and occasionally enter them. Much of their energy seems to be profitless. They stalk and spring on one another. They also molest the females.'

The females do not always seem to be as responsive to their overtures as the males might wish. They are busy matrons with a purpose in life. There is a time and a place, they seem to feel, and it is irritating to be pestered with amorous proposals when there is serious work on hand. The female wasp, in fact, has a short way with suitors who accost her when she is returning to her nest with prey. She avoids them, she ignores them, she makes it clear that she has no time for them. The passionate males, however, cannot bring themselves to give up hope and they have been seen occasionally to cling to the backs of females and so prevent them from entering their burrows. The

resourceful female thus embarrassed has no difficulty in dealing with her suitor—she simply scrapes
him off by walking between two pine needles.

The males, however, have evolved a means of
having their way with their unenthusiastic mates.
Fabre has described how he settled down to watch
some wasps flying about in a sandy clearing. He
noticed that there was a large number of them
and it was clear from their small size and their
characteristic flight that they were males. They
flew very low, almost grazing the ground, and
seemed to be passing and re-passing over the same
area. Occasionally one of them would settle on the
ground and feel it carefully with his antennae. He
had the air of trying to find out what was going
on underneath. Then he would begin to fly again,
to and fro, backwards and forwards.

Fabre was intrigued by their behaviour. They
seemed to be looking for something, but it was not
clear what it might be. It did not seem to be food,
for there were plenty of flowers close at hand of a
kind on which wasps normally like to feed; yet none
of them settled on the flowers or drank their nectar.
Their whole interest seemed to be concentrated on
the ground.

The explanation was that they were waiting for
females to emerge from it, newly-born females
who had that moment burst from the cocoon and
forced their way upwards through the sand to the
sunlight. 'She will not be given time to brush herself or to wash her eyes,' says Fabre of the emerging

female. 'Three or four or more of them will be there at once, eager to dispute her possession.' The males, it seems, generally hatch from the cocoon rather earlier than the females and they make the most of this temporary advantage by lying in wait for the females to emerge and mating with them when they are momentarily a little dazed and bewildered by their strenuous experience and their first sight of the bright world above ground.

The female wasp is not always indifferent, however, and the males need not always resort to guile in order to win her. There are times when she can be tempted.

It is a charming scene, the courting flight of some solitary wasps. The splendid females fly gravely humming in the sunshine, flitting from bush to bush. The males, who are ever on the watch for such an opportunity, come swiftly after them. They join the happy swirl, and soon couples are formed, and parted, and joined together again. The course of love may not always proceed without an ugly scene, however. Sometimes a rival male may bustle up to one of these couples and the peaceful humming changes to an angry note. Battle is joined and the two warriors roll in the dust. The female, demure—or indifferent—quietly awaits the finish of the fight. At last the triumphant victor rises in the air and seeks out his formidable prize. She is not reluctant. The united couple fly away together, away from the crowd, away into the quiet and solitude of some distant brushwood.

'Here,' remarks Fabre, 'the part played by the male ends.' He is unjust, for he forgets the *Trypoxylon* male, a shining example to his pleasure-loving confrères. It is true that he makes no attempt to undertake any of the heavy work of domestic life—his physique would scarcely permit of it—but he sets up house with his mate and helps her as well as he may.

When the female has carried out the arduous business of making the nest, she goes out hunting and the little male takes up his station just inside the nest, his head very nearly filling the entrance. He stands guard in this position, only playing truant occasionally to take a short flight, until the nest is finally stored and closed up. He has enemies to guard against. There are parasitic flies which are always on the look-out to creep into the nest, and these he drives away with splendid vigour. Unattached males are a threat, too, and they may try to creep into an unguarded nest. These are nearly always defeated, however, by the resident male, who is no doubt fortified by his righteous sense of possession. It is a fact among the lower animals that right often appears to confer might— birds, for instance, are nearly always successful in defending their little piece of territory from invaders even when the invaders are stronger and larger.

It must be confessed, however, that the *Trypoxylon* male falters in one respect. As so often in the insect world, all is not quite so idyllic as it might seem, for when the invader of the nest is a 'strange female' who has perhaps mistaken the nest he is guarding for her own, he makes no objection to her entrance—indeed, he might even be said to welcome her in. His mate, however, is seldom in a position of moral superiority, for the females, too, have been observed to be not at all put out if a strange male presents himself at the nest at a time when the rightful owner is temporarily away. But as soon as the owner returns, he drives out the intruder and domestic regularity is restored.

Apart from his indulgent attitude to strange females, however, the male *Trypoxylon* shows himself a valiant guardian of the nest. The Peckhams sometimes teased these males by pushing a blade of grass at them, but they never gave way— they always attacked it with great spirit and some-times gripped it so fast with their mandibles that they could be pulled right out of the nest on the end of it. The male *Trypoxylon* is sometimes helpful in other ways, too. When the female comes back to the nest with her prey, he comes out to make way for her, then jumps on her back as she goes in. He comes out again in the same way, but leaves her at the entrance and takes up his position on guard. Occasionally, so the Peckhams say, he does better. They relate that he sometimes relieves the female of her spiders as she brings them in and packs them in the nest himself, so leaving her more time to spend on hunting.

The domestic virtues of the *Trypoxylon* male are exceptional, but even if they were common to all male wasps, they are no great matter. In assessing the attractions of this insect, they could hardly be held to counter-balance the extreme cruelty of the means by which the female provides for her young. And yet it is this macabre creature which seems, above almost all others, to inspire affection in human beings. The English naturalist Lubbock seems even to have made a pet of one.

I took her with her nest in the Pyrenees, early in May (he writes in his book *Ants, Bees and Wasps*). I had no difficulty in inducing her to feed on my hand; but at first she was shy and nervous. She kept her sting in constant readiness; and once or twice on the train, when the railway officials came for tickets, and I was compelled to hurry her back into her bottle, she stung me slightly—I think, however, entirely from fright.

Gradually she became quite used to me and when I took her on my hand, apparently expected to be fed. She even allowed me to stroke her without any appearance of fear, and for some months I never saw her sting.

When the cold weather came on, she fell into a drowsy state, and I began to hope she would hibernate and survive the winter. I kept her in a dark place, but watched her carefully, and fed her if ever she seemed at all restless.

She came out occasionally, and seemed as well as usual till near the end of February when one day I observed she had nearly lost the use of her antennae, though the rest of the body was as usual. She would take no food. Next day I tried again to feed her; but the head seemed dead. The following day I offered her food for the last time; but both head and thorax were dead or paralysed; she could but move her tail, a last token, as I could almost fancy, of gratitude and affection. As far as I could judge, her death was quite painless; and she now occupies a place in the British Museum.

What is the secret of the wasp's charm? She is pretty and graceful. She is a hard-working and conscientious mother—that is a virtue which none can deny her. She is brave, and if need be she will attack fierce insects larger than herself. She carries out her work of slaughter with clockwork efficiency, and she may plead that she does not kill for herself—she kills that her children may live.

And as for the dark horror of that underground feast on a living victim, what does she know of it? She has sealed the nest and gone her way. Most likely she will be dead long before her child emerges to the sunlight and, in its turn, the puppet of implacable instinct, slaughters for the coming generation which it, too, will never see.

THE SOLITARY BEE

1

PUPPETS OF INSTINCT

LIKE the solitary wasp, the solitary bee lives the greater part of her life quite alone and has none to help her in the labour of providing for her children; but at least they *are* her children. Unlike the worker-bee of the hives, she is able to perform the functions of her sex in the fullest sense and is not reduced to the status of a virtually sexless creature who must labour her life long to provide for the children of another.

The life of the hive is indeed singularly repellent to the human observer. The solitary bee is without the support of a band of fellow-workers, she knows nothing of the shelter of a hive, but her pioneering life, with all its perils, is surely preferable to that of the sterile workers, the doomed and helpless drones and the grotesquely prolific queen of the hives. The female solitary bee is no idler—the whole of her short life is given over to the fight to preserve her race. She must construct her nests, she must provision her nurseries and she must lay her eggs. But she carries out these tasks with ingenuity and skill and perhaps she may find some satisfaction in the doing of them.

Whether she does or not, however, it would be a mistake to be too much impressed by a show of intelligence, for when it is put under strain, she often seems curiously obtuse. None the less, so long as her life follows its accustomed rhythm, she does well enough and shows quite astonishing aptitudes in certain directions. The different species of solitary bee have their own traditional methods of constructing their nests and have developed their own special skills. The leaf-cutter bee is skilled in cutting ovals and circles out of the leaves of certain trees and shrubs and plants so that she may line her nests with them. This rather elaborate form of upholstery makes sufficient demands on her time; to dig or construct her nests in addition would call for more effort than she could reasonably give to this part of her work and would leave her no time for providing for her family. So most leaf-cutter bees first take possession of a ready-made structure—a reed stump, perhaps, or the discarded nest of another kind of solitary bee, or sometimes the burrow of a large earth-worm. The next task is to cut out her leaf-snippets by standing on the top of a leaf and gnawing a piece away with her mandibles. The first pieces she cuts are large and irregular, and she pushes them down into the bottom of her burrow, if she is using a hole in the ground. These seem designed to play the part of a fortification to keep out unwelcome intruders from below. Next she cuts fairly large oval pieces which she uses to line the bottom and sides of the cell which is to contain her egg. Then she cuts a number of smaller pieces to fill up the gaps and give greater thickness to the whole.

At this stage, there is an interruption in building operations, for the mother must now set about laying in the provisions which are the universal fare of the young bee. The bee larva is a vegetarian like its mother. Unlike the wasp larva, it demands no gruesome feast on a living but helpless victim, which must die slowly hour by hour so that the larva may grow fat. The young of the bee is content with honey and pollen. No doubt as a result of this fact, bees have developed a peculiarly intimate relationship with flowers and their mouth-parts have become modified so that they are well equipped to lap up nectar. In fairly primitive bees the mouth-parts are short, but in more specialised forms the 'tongue' has a channel on its under-surface covered over by special hairs so as to form a sort of tube. The nectar is drawn up along this passage and is stored in the crop. Here it is mixed with an enzyme, which probably comes from the salivary glands and has the effect of converting the sugar of the nectar into another form of sugar—into honey which the bee can regurgitate to feed her young. Bees have also developed special modifications for collecting pollen from the flowers. The whole body of most bees is covered with branched and feathery hairs to which the pollen clings and by these means the bees carry pollen as well as nectar to their nests and can thus provision their cells with a mixture of the two. On the other hand, the parasitic bees which feed their young on provisions collected by the more industrious species, and the male bees which take no part at all in collecting food for the young, have none of these special adaptations.

When the leaf-cutting bee has gathered in her provisions and laid her egg, her next task is to close up the cell; and here, according to Fabre, she shows gifts which seem truly astonishing. She flies away from her cell, which is nearly always in darkness— if it is built in the burrow of an earth-worm, the darkness must sometimes be almost complete. Yet in her mind, it may be by feel—it can scarcely be by sight—she carries an impression of the exact size of this cell. For she flies to a leaf and very frequently, though not quite invariably, she cuts out a round which fits exactly and precisely over the mouth of the cell. This seems a remarkable feat. The explanation may be that the size of the cell and the size of the leaf fragments are both determined by the size of the bee, which uses her legs and body like a compass to draw a circle of the right size. Even if this less romantic explanation is the true one, the accomplishment still seems impressive enough. This insect, responding, as it must, to the inexorable demands of evolution, has developed the special talents it must have in order to survive. This talent she needed, and she has won it.

She carries back the circular piece of leaf and

fits it over the cell. Then she flies away and cuts another, and another—there may be as many as ten pieces placed over each cell. The first few pieces are nearly always of the same size and nearly always fit the opening exactly. The later pieces may be slightly larger and pressed down so as to turn up a little at the sides and form the curved floor of the next cell.

The bee builds a row of these cells—generally about five or six, although a dozen or more is not unknown. When the cells are completed, she fills up the entrance to the burrow with a large number of pieces of leaf which she cuts more or less at random and without much regard to size or shape. Here there is no need for the delicate precision which she exercised in covering the cells. Quantity is what she needs to make the barrier secure, and she sets about her work with as much speed as possible.

The summer is passed in this labour of cutting leaves to line her cells, provisioning them with honey and pollen, and laying her eggs. But in time her ovaries become exhausted and the season of her fruitfulness passes. She could rest now and none would blame her. She could sit in the sun and drink nectar from the flowers for her own enjoyment. But she is set in the habits of a lifetime. She has carried out her routine of labour all her days and now she is its prisoner. She selects a burrow, as she has always done, and she begins to fill it with leaves. She goes on and on with her

futile task until her burrow is filled with literally
hundreds of pieces of leaves. She fills it to the very
top, until it overflows. A formidable barrier is
built, and there is nothing to protect. She has no
eggs to lay and she has lost the urge to bring in
provisions for young ones which will now never
be born. She is, in truth, old and worn out. But
all her life she has been used to cut leaves and store
the pieces in holes, and now she cannot stop.

It might perhaps be argued that she shrinks from
admitting the fact that she has become useless and
strives to hide the truth from herself by this
semblance of purposeful activity; but that would
be to credit her with thoughts she can scarcely
possess. The truth seems to be that she is the victim
of blind urges. She was born and she was mated,
the wheels of her life's work were set in motion;
and only death can bring them to a stop.

The same futile persistence in a useless task can
be seen in other kinds of solitary bee. The Three-
horned *Osmia*, like the leaf-cutter bee, adapts a
ready-made nest to her use and often chooses an
empty snail-shell for this purpose. She divides
up this dwelling into little cells walled off from each
other with dried mud, provisions each with honey
and lays an egg in it, and then fills up the entrance
to the shell with a stout and massive plug. This
is her life's work, endlessly repeated. But the time
comes when her ovaries are exhausted and her
useful work has come to an end. Like the leaf-
cutter bee, she cannot recognise her fate or accept

it. She places a few specks of pollen in an empty shell and then devotes hours to closing up this unimportant treasure. Or she may build a few partitions inside a shell and then abandon the nest. On and on she goes with her unproductive labours until at last she dies. As Fabre has said, 'This worker knows no rest but death.'

The limitations of instinct do not show themselves only when the machine of life has begun to run down—they are liable to become apparent at any time when the known and accustomed rhythm of life is interrupted. For instance, there are certain kinds of solitary bee which spend their energies more on the actual construction of a nest than on elaborate linings to the cells. Some of these are Mason-bees which live a sort of village life by settling down in fairly large colonies, perhaps under the tiles of a house or a balcony. Other Mason-bees seem to prefer to go off on their own and construct their nests on a bare flat stone, or on a branch or twig. Both kinds make use of the same material for their building; they use clay or chalk, often scraped up from a road, mix it perhaps with a little sand and then moisten it with saliva. This forms a paste which eventually hardens into a material of a consistency not unlike cement. They carry this material to their home in little pellets, work it into shape with their fore-legs and mandibles and strengthen it with small pieces of gravel.

Methods of building vary slightly with the different species, but the Mason-bee which builds on flat surfaces constructs a cell about two centimetres high. She uses small pebbles or fragments of gravel to form the walls and binds them together with her mortar. The outside of the cell is left fairly rough, but she gives the inside a complete coating of mortar and spends time and effort in working it to a reasonably smooth finish.

The bee then sets about stocking the cell with food and journeys busily to and fro carrying honey and pollen. Every now and again, when she has accumulated a fair store, she pauses to give the mixture a vigorous stirring with her mandibles and so form it into a paste.

When the cell is about half full, she judges the supply of food to be sufficient and lays her egg on top of the paste. Her last task is to close up the cell and this she does with considerable care, using mortar unmixed with any other material. All this takes about two days and, when the work is completed, she proceeds at once to the construction of the second cell, which is built adjoining the first. In all, clusters of about six or eight adjoining cells are constructed, provisioned and equipped with an egg. Then the bee proceeds to make the cluster of cells safe against the extremes of heat and cold which they might suffer in their exposed position, and also against the attacks of enemies and the innumerable accidents which might threaten her family. Carrying her mortar in her mandibles, pellet by pellet, she proceeds to cover over the whole structure to a depth of about a centimetre. In the end, a mound is formed about the size of half an apple.

The mortar sets almost as hard as a stone and it might be thought a difficult matter for the tender and newly hatched bees to pierce this formidable barrier. They are equipped by nature for this task, however, and apparently perform it without difficulty by biting through the hard clay with their mandibles. Réaumur relates, however, that some emerging bees were vanquished when faced with the further barrier of a little gauze. A friend of his placed a nest in a glass funnel and covered the top of the funnel with gauze. The bees—three males— duly emerged from their nest, but after that made no further effort to achieve freedom. They died as prisoners in the funnel, either because they were unable to penetrate the gauze or because the idea of attempting to do so never occurred to them.

Fabre was dissatisfied with this experiment. The whole thing seemed to him to be ill-conceived. If the idea was to find out whether the bees would attack a second obstacle after successfully negotiating the first, gauze was not a very suitable material from which to form it. The bees are equipped by nature to deal with hard mortar and their tools might well be unsuitable to tackling gauze. Also it seemed to him a mistake to allow the insect to find itself in broad daylight before the second barrier was pierced. It would be distracted, he thought, by the sight of the sunshine outside and would wear itself out by beating itself against the glass. To expect it to find out that the way to freedom lay through the comparatively opaque gauze was, he thought, to demand reasoning powers beyond the scope of an insect.

Fabre decided, therefore, to present some bees with thick brown paper as a second barrier. First, however, he wanted to find out whether they had adequate tools to penetrate the paper, so he placed some cocoons in the hollow stump of a reed and covered over the openings with either clay, sorghum or brown paper. All these barriers were overcome apparently with equal ease and in identically the same manner—a neat round hole was cut in each, exactly similar to the hole by which the bee escapes from her nest in the ordinary way.

Fabre had also taken two nests which were intact and still resting on their pebbles. Over the top of one of these he pasted a piece of brown paper, so that although the bee was faced by a double barrier, there was no space between the two and to the bee it might appear like a single barrier of unusual difficulty. On the top of the other nest he glued a brown paper cone, so that although the emerging bee was faced with only the same amount of material to penetrate, it would seem like two separate barriers instead of one.

The bees in the first nest appeared not at all disconcerted by the addition of the brown paper to their usual covering. They emerged in good time and apparently without difficulty by punching a hole in the brown paper as they had already done in the hard mortar. The fate of the bees in the other nest was pathetically different. They duly bit their way out through the mortar, and then found themselves in darkness, in a sort of tent. Such a predicament was quite outside the experience of their kind and nothing in the past history of their race had prepared them for it. All that debarred them from complete freedom was one layer of brown paper. With their powerful mandibles they had the means to bite a hole through it without the slightest difficulty—that fact had already been demonstrated with others of their kind. The obstacle was flimsy. It was negligible for such well-armed insects; but they did nothing. With freedom so near, so easy to obtain, they died in their paper prison.

Again it is the limitations of instinct which made them helpless. When the moment to emerge from the nest arrives, certain processes are put into action which move the insect to attack the barrier which confines it and escape into the air. When this act is once accomplished, the urge has reached its natural conclusion. It is finished and over, and there was apparently nothing in the new conditions in which these insects found themselves to rouse them to repeat the action a second time. They could crawl and move about in their new prison and that represented to them, as Fabre says, 'the end of the act of boring.' In order to escape, the insect would have had to repeat an act which, in the ordinary course of nature, it is only called on to perform once in all its life. The bee cannot reason, it is not free to go back a stage in the cycle of instinctive behaviour, it has not the intelligence to adapt itself to new circumstances. The instinct which serves it so well in the ordinary, the expected, problems of daily life, now become its murderer.

Fabre carried out other experiments which demonstrated the same lack of adaptability in the bee. He moved a pebble on which a nest had been built about two yards away from its original position and awaited the return of the owner. In a few minutes she came and made straight for the place where her nest had been. She hovered over this place, settled on it, walked about on it, all the time searching with passionate concentration. Her nest was clearly in view, only a short distance away, but she took no notice of it. At last she flew away; but soon she was back again to resume the search. These manœuvres were repeated many times—a search, a flight away in apparent exasperation, a return and a renewed search. Often she passed directly over her nest in its new position, sometimes she was only an inch or so away from it; but she ignored it. To her it was no longer her own.

Sometimes, if Fabre moved the nest only a yard or less from its original position, the searching bee would examine it carefully and even dip her head inside it to inspect the provisions which she herself had lately gathered. But even so, she was apparently quite unable to decide that the nest was really her own property, standing, as it did, in this new and strange situation. Always she abandoned it and set herself to building afresh.

The result was different, however, if Fabre removed her own nest and offered her another bee's nest in its place. The new nest was accepted readily and apparently without question. Fabre was careful, however, to offer her a nest in the same stage of development as her own so that she could continue the work on which she was already occupied. 'If she was building,' he writes, 'I offer her a cell in process of building. She continues the masonry with the same care and the same zeal as if the work already done were her own work. If she was fetching honey and pollen, I offer her a partly provisioned cell. She continues her journeys . . . to finish filling another's warehouse.'

An exchange back to her own cell was equally unremarked by the bee. Fabre was able to carry out a series of such exchanges, so that the bee worked alternately at her own cell and at another's, without at any time showing any sign of suspicion or disquiet. The one essential condition was that the new nest should always be at exactly the same stage of development as that which was taken from her. Similarity in appearance and design was not essential at all. Fabre substituted nests of vastly different appearances and the bee was still unperturbed. All she demanded, apparently, was that she should find a nest in the place she had learnt to recognise and that she should be able to continue work on it without any need for a change of plan.

It seems that the bee is able to remember a locality and find her way to it. Fabre carried out experiments with bees similar to those he made with the solitary wasp. He carried bees in paper bags by devious routes to places several miles from their nests and was able to show that the majority succeeded in making their way home. His experiments were unfortunately not designed in such a way as to give clear proof that bees have a homing instinct similar to that possessed by some birds—his bees may well have flown as far afield before and by luck or persistence they may have managed to find landmarks which guided them home. But however that may be, the majority *did* find their way. And when once they are in the neighbourhood of their nests, bees are generally able to locate them with precision. Most of the experiments which have been carried out in recent years seem to indicate that they do this chiefly by sight—that they memorise and recognise certain landmarks or conspicuous objects which lie near the nest.

But of the nest itself they seem to have no clear picture. Why should they? Animals develop, in the course of evolution, those faculties which they need in order to survive. If the race of solitary bees is to survive, it is essential that the mother bee should be able to find her way to and from her nest so that she can carry out the work of building and provisioning it. An ability to identify the nest itself would be superfluous so long as she remembers its exact position. Evolution has not equipped her to deal with the interfering devices of naturalists, and short of such interference it seems highly unlikely that she would ever have to cope with the worry of a nest which transported itself in her absence two yards away from where it had lain before. It is less likely still that she would ever meet in nature the strange phenomenon of another's nest standing in exactly the position where she had, but a few minutes before, left her own. An ability to cope with such eccentric happenings would be of little use to her in the ordinary affairs of life.

Fabre next made the experiment of offering a bee a cell which was in a different stage of development from her own. He took away a cell which was still in its very first stages and put in its place a completed cell almost filled with honey. It is true that such a splendid windfall would be hardly likely to perplex the bee in nature, but even so it does not appear to demand a very high degree of intelligence for the bee to be able to take advantage of it. Even the most dull-witted bee must surely notice that here before her is a completed cell, and all that is now needed is to bring in a little more honey and pollen, lay an egg in it and close it up.

Fabre's bee did no such thing. The cycle of the bee's instincts unfolds itself rhythmically and inevitably. All things must be done in due order and in due season. She was building and build she must. The only concession she was able to make to the unexpected was that she built a little less than if she had been working on her own cell. She built up the completed cell until it stood about a third as high again as is normal, but no more. After building comes provisioning, and she set out to lay in her store of honey and pollen. But here again she made her small concession to a change in circumstances. Adding her provisions to those which were already there, she built up a stock far in excess of normal needs, but none the less she did not bring in quite so much as if she had been provisioning an empty cell.

The same thing happened when Fabre did the opposite and gave a bee which had passed the building stage an uncompleted cell. The bee, returning with her crop full of honey, did indeed seem a little perplexed as to where she was to put it. She considered the new cell carefully and felt it over with her antennae. She flew away, came back, flew away again. Fabre hoped that she would jettison her honey on one of these journeys and come back with building materials. But it was impossible for her to do so. The time for building had gone by, the urge to provision a cell was upon her and she had no choice but to obey it. Rather than rebuild her own cell she would prefer to disgorge her honey in another's. And this, in fact, was what the bee occasionally did.

Fabre tried slightly different experiments to test the bee's ability to adapt herself to changed circumstances. For instance, when one bee was in the middle of putting a covering over her cell, Fabre made a hole in it while she was away. This did not disconcert her at all. When she returned, she apparently noted the damage, repaired it with skill and finished the covering. She was already engaged on building and Fabre had set her a building problem. He then pricked a hole in the bottom of a cell which a bee was constructing. Again she remarked the damage and repaired it promptly.

Fabre next made a hole in the bottom of a cell which a bee was engaged in provisioning. The honey gradually drained away, but the bee took no notice. She laid her egg on the diminishing store of food and sealed up the cell with exemplary care. Fabre tried the same experiment again, this time choosing a cell in which the first stages of building had just been completed. The bee returned with her load of food which immediately ran out through the hole. She was not unaware of what was happening to judge from her actions. She put her head into the cell, felt the hole with her antennae and carefully explored it. Then she flew away, but not to return with mortar for repairs. She came back with more provisions. She came back with more and more and more, desperately trying to fill up her cell which emptied itself inexorably. In between her bouts of provisioning, she turned to building work, as is the custom, adding fresh storeys to the cell; but still she did not plug the hole in the bottom. That belonged to a stage in the work which was past. At last she had completed her urge to build and to provision, and she laid her egg on the thin ooze of honey lying at the bottom of the cell and then closed it up as usual. Even though her child would starve, she could by no means go back and attend to a part of the building which had once been completed, although a breach in that part on which she was engaged would be attended to at once.

Fabre demonstrated this again when he made a hole in the wall of a cell at a time when the bee had almost completed bringing in provisions. She laid her egg and proceeded to seal the cell with meticulous care, stopping up and mending every minutest crack or fault in the lid. Yet all the time a large window was gaping in the side. She did not ignore it—she did, in fact, seem a little troubled by it, for she went to it over and over again, examined it carefully, felt it with her antennae and tentatively nibbled the edge. But to repair it was beyond her—the building of a wall belonged to a past which she had no power to recall, and in the end she went away and left her egg exposed to any enemy which might care to enter through the breach.

Even in matters of small importance the bee shows this strange lack of elasticity. It is her habit when she arrives at the cell laden with honey and pollen to deposit the honey, then to emerge and go in again backwards to brush off the pollen. Fabre kept interrupting a bee at the moment when she was about to enter backwards by pushing her on one side with a straw. Each time the bee went back to the beginning again, entered the cell head first, although she had no honey to deposit, and then tried to go in backwards. Fabre interrupted her at this moment over and over again and she could never learn to go on and complete the action from the stage at which it had been interrupted. One is reminded of the unfortunate people who suffer from obsessions and often develop complicated rituals which they must carry out in exact order as they dress or wash or settle to their work. The ritual must always be the same and its performance only satisfies them if it is carried through smoothly and without a break. As for the bee, nobody can tell whether she feels satisfaction or not in the performance of her tasks; but perform them she must, and it seems she has little freedom to interrupt or change or vary them. The machinery is set in motion and the wheels must turn in their appointed rhythm.

2

TENANTS AND PARASITES

THE solitary bee devotes painstaking care to the construction and closing-up of her cell, but even so she cannot fully protect her young from the parasites which prey on them. Indeed she often seems quite unaware of danger even when it is paraded before her without concealment. Like the solitary wasp, it is often within her power to destroy the devourers of her children, but she lets them live.

The Anthrax fly, however, attacks with a subtlety and determination which might well defeat the most exemplary vigilance. This fly lays her eggs near the nests of those Mason-bees which seal their cells with a kind of mortar. She takes no special care where she deposits them, she does not cover them with sand—she merely drops them so that they lie on the ground in the glare of the hot sun. Quite soon, a very tiny creature, no thicker than a hair, but active and wriggling, emerges from the egg. It knows what it must do. Without delay it makes for the bees' nests and searches persistently and unremittingly for some tiny crack in the mortar through which it may creep inside. It has only a limited time in which to succeed, for if the growing bee passes beyond a certain stage of development it will no longer provide suitable nourishment. The grub's mission, then, is urgent and involves incredible difficulties for so weak a creature. It is helped by one thing— it is able to continue without food for a considerable period. This is just as well, for it must often search a long time in vain before it finds a crack at all, and often it must follow false trails which lead nowhere and which it must later abandon. The best it may hope for is to wriggle on through a maze of tiny cracks until at last, after much effort, it is inside the bee's cell.

And here it must exercise patience. It is not always possible for it to fall on to its meal without delay and break its long fast at once. Whether patience of a conscious kind is called for, or whether the small creature is blindly obedient to the stimuli which prompt its actions, it must at all events wait to eat until the bee larva has begun to grow quiet and lethargic in preparation for its change into a pupa. This is the moment when the Anthrax larva establishes itself on its future host and actively reconnoitres the meal that is to come. Drawing itself up into loops like a caterpillar, it travels about, sometimes rearing itself up and seeming to search the air with its tiny, amber-coloured head.

The time of the banquet at last arrives as the bee larva changes to a pupa and passes into the curious state of suspended life in which it sleeps

out this stage of its development. It is a sleep
from which it will never wake, for now the
Anthrax larva begins to feed with zest. Its fast
has been long and its appetite is lusty. It quickly
transforms itself into a smooth fat grub equipped
not with mandibles or any instrument for piercing
the skin of its host, but with a small mouth like a
sucker.

The creature is not obliged to proceed with its
meal after the manner of a solitary wasp larva
devouring its paralysed prey. The wasp larva must
avoid killing its host too soon, and this it knows how
to do, provided that it is not disturbed. If it is once
disturbed and is thus obliged to make a new beginning, it finds it has lost its bearings and it goes
astray on its course. It may not begin to feed again
at all, but, if it does, it is likely to kill its host by an
injudicious bite and poison itself with putrefying
fare. The Anthrax larva has none of these problems.
It can break off its meal at will. It can apply its
mouth now in this place, now in that—it makes no
difference, for the creature does not bite; it sucks.
Somehow it manages to draw out through the
unbroken skin the whole substance of its plump
and once healthy host.

At first, there is little to notice. The Anthrax
larva is perhaps a little fatter, the bee larva develops
a faintly dejected look. The glow of full health is
beginning to fade. Soon the process goes a little
faster. As the Anthrax larva grows stronger, its
appetite grows with it. The bee larva shrivels,

and folds appear in her skin. She grows slack and flaccid. But the Anthrax larva is doing well. It is now as fat as butter, sleek and shiny with good living. At last its host is reduced to a mere empty bag of skin, dead and dry. By her side, the Anthrax larva, replete and satisfied, prepares to pupate.

It sleeps and develops, until in due season it undergoes certain changes. This is necessary. It is enclosed in a cell of hard mortar. It is not endowed with the powerful mandibles which enable the bee, the rightful occupant, to escape. It is far too fat to think of wriggling its way through the minute fissures by which it entered. The processes of evolution have not failed, however, to endow it with the tools for the job. It grows six strong spikes on its head in a semicircle, like a sort of coronet. This armament it uses as a battering ram. It also grows spikes on its abdomen, and a good supply of stiff bristles all over its body, which serve to give it a purchase on the wall and hold it up as the work proceeds.

Even with this equipment, the matter of forcing an exit is not simple. The little creature draws itself back and then allows its spiky head to strike against the mortar. It climbs up and batters again at the barrier. At last, after much effort, a hole is forced into the outer air and the head and thorax appear in the opening. This is the moment for its last and final transformation. The head splits open, a rent appears in the thorax and out through this opening emerges the adult fly. For a moment she seems dazzled and confused. She waits a little while the sun dries her newly-opened wings. Soon she spreads them and flies away to enjoy the few weeks of freedom which now lie before her—the brief life in the sunshine which she has won by such arduous and such protracted labours. She will feed on the flowers and mate with her kind. She will scatter her eggs on the dry sand; and soon she will die. But her eggs survive and the cycle goes on.

Mason-bees suffer from the attacks of other parasites, too. The hard coverings of their nests seem surprisingly ineffective, in fact. A parasitic bee called a *Stelis* manages to gnaw her way through into the bee's nest, although she is neither large nor powerful and the undertaking seems arduous in the extreme. She gnaws first through the outer covering, then through the lid of the cell and at last she reaches the honey. She does no harm to the bee's egg which is already there, but lays her own eggs beside it. The eggs develop into larvae with big jaws which devour the egg of the bee. After this precaution, they lose their jaws and feed peaceably on the food which had been stored by the bee for her own offspring.

The *Stelis* has not finished her work, however, when she has laid her eggs. If she is disregardful of the bee's welfare, she cares for her own young and will not leave them to grow up in a nest open and exposed to any passing thief. She collects a little earth, moistens it with saliva and patches up the hole. Fabre writes that in his district the *Stelis* scrapes earth from the foot of the pebble, and this earth is generally red, whereas the bee has built her nest with powder scraped from flint. A red patch, then, standing out against the near-white of a bee's nest, is the mark of the invader.

A certain kind of fly, the *Leucospis*, is also a violator of the Mason-bee's nests. She is armed with a long sharp probe, which serves a function similar to that of the needle of a hypodermic syringe. She manages, by considerable effort, to drive it through the hard cement. This may take her a mere quarter of an hour, if she is both lucky and skilful. At other times she may have to labour hard for a good three hours before her needle at last penetrates to the honey. Then she is free to deposit her egg from which will emerge, in due course, a hungry larva, bent not upon sharing the bee larva's rations, like the *Stelis*, but upon using the bee larva itself as a ration.

A buttery, shiny larva hatches out and sets upon its host. Like the Anthrax grub, it appears to drain its host dry without ever breaking the skin until at last, fat, satisfied and splendidly replete, it settles down to complete its cycle of growth by the side of the empty bag of skin which is all that remains of the bee larva. The skin has been penetrated, however, for the *Leucospis* larva is equipped with mandibles of great fineness and delicacy. They are of no use for the rough work of biting and chewing, but they serve well enough to pierce the skin of its host and make a tiny breach through which the nourishing fluid may be drawn up.

Bees may also suffer from the attacks of their own kind. The *Nomada* bees, sometimes called the Cuckoo bees, provide for the future of their families by taking advantage of the labours of their fellow-bees. They are brightly-coloured creatures—rust-red with black stripes and yellow spots. They do not labour as other bees do—they lack even the implements of labour. They have no pollen-baskets, their bodies are smooth and shiny and bear no hairs to which pollen might cling. This is no disadvantage to them for they have no need to gather pollen—others will do that for them. They wait until they find a cell ready-built and stocked with food, and then they enter it and lay their eggs inside. What could be simpler or more desirable? Particularly when the rightful owner has apparently no objection to this programme.

Sometimes it happens that the owner of the nest comes home bearing food while the *Nomada* bee is inside and occupied in carrying out her nefarious purposes. 'When such proved to be the case,' writes Mr. F. Smith, 'she would issue from it, and, flying off to a short distance, wait patiently until the parasite came forth, when she would re-enter and deposit her burden.'

It used to be thought that if the owner returned and found an egg ready laid in her cell, she would conclude that she must have laid it herself and absentmindedly forgotten the event; or, to put it rather differently, the sight of the egg apparently provided a stimulus which induced the reaction of shutting up the cell and going off to build another. It seems more likely, however, that the bee proceeds to lay her egg as usual, but the parasite emerges first and eats up the honey, so bringing about the death of the rightful occupant.

There seems to be some doubt whether these *Nomada* bees are rightly called parasitic. 'Some authorities consider that the term inquilines (Lat. *inquilinus*, a tenant) better indicates their relations with their hosts,' writes the British entomologist, A. D. Imms. This view is apparently based on the fact that although these bees are highly injurious to their hosts and bring about their deaths, the owner of the property has no suspicion of their designs and in consequence tolerates their presence. It seems a pessimistic view of the habits of tenants.

The solitary bees seem exposed to the attacks of a multitude of enemies. The *Tachina* is their enemy, as she is the enemy of the solitary wasp. The *Anthophora* bee, which makes its nest in old walls and chalk pits, is preyed on by earwigs, the *Melecta* bee and the Firetail-fly or Ruby-tailed wasp. This last creature is capable of showing what must seem to us a dazzling courage in obeying her urge to provide for her family. A French naturalist has described how he saw one of these insects enter the nest of a bee while the owner was away. The bee returned with her load of pollen and instantly attacked the intruder with savage violence. The Firetail-fly rolled herself up into a ball, after the manner of her kind, but the bee persisted in her attacks and eventually succeeded in biting off her opponent's wings. Then she dragged her victim to the nest entrance and threw her outside. No doubt feeling that she had dealt with the situation adequately, the bee then deposited her pollen and flew away on her business.

The wretched, mutilated, half-dead Firetail, scarcely capable of movement at all, none the less dragged herself up the wall to the bee's nest, some-how managed to crawl inside and there laid her egg. Even with this she was not satisfied. She spent the last of her strength in pushing her egg with great care between the pollen and the wall so that it should not be visible to the bee when she returned.

The *Melecta* bee, a handsome creature with its jet-black head and thorax and its silver tufts on the abdomen, preys on her sister bees, but suffers a just fate in that she herself is the victim of the *Meloë* beetle. The *Halictus* bee, which has developed a very rudimentary kind of social life, is also victimised by parasites which it persists in treating with foolish good humour. This bee forms her nest by burrowing in firm soil and making a gallery about ten inches long from which small and finely-finished apartments lead off. It is in these that she establishes her young and she equips each one of them with a firm ball of a paste-like substance kneaded from honey and pollen. It is not surprising that other insects attempt to secure the use of these desirable nurseries for their own families.

One of these pirates is a fly—a tiny creature with dark red eyes set in a white face, pearl-grey body and black legs. She settles herself among the nests of the *Zebra Halicti*, which are built close together to form a kind of village, and patiently waits until she sees a bee returning with her load of honey and pollen. Then she sweeps up into the air and keeps close behind the bee as she flies hesitantly from side to side, looking for the landmarks by which she may recognise her own nest among the huddle of other nests.

At last the bee finds her own doorstep and disappears inside. The fly drops down close beside the nest and once again settles down to wait, her eyes fixed in an unmoving stare on the entrance. The owner reappears and stands for a moment on the threshold. The fly is not embarrassed. She makes no move to go away or to conceal her presence. The two insects may be an inch or less apart, but the fly shows no fear. Nor does the bee make any sign that she is angry or alarmed for her family's welfare. Complete indifference, as Fabre has said, marks the encounter on both sides.

The bee flies away. Nothing could have been easier for her than to destroy the puny murderer lying in wait at her door—as an adversary she would have presented no problem at all. But she ignores her and flies away. The opportunity is gone and one, at any rate, of her children will pay the price of her negligence.

The fly walks into the nest with an air of leisure —it is as though she knows the bee must be away for some considerable time gathering her stock of honey and pollen. She enters the cell and lays her eggs on the cake designed for the young of the bee and then goes out in as leisurely a manner as she came in. Sometimes, however, she finds the cake is unfinished and not ready to receive her eggs, so she withdraws and waits outside again. She may even, with incredible daring, accompany the bee on her journey down into the nest in order to inspect the work as it progresses. Even this liberty the bee appears to suffer without rancour—at any rate, the fly emerges from the nest with no sign of flurry or anxious haste.

The disaster she produces for the bee's young, however, should soon be obvious to the bee. The cells are not sealed off until the bee's larvae are ready to pupate and at any time during the interval the bee may look into the cell and observe the slow starvation of one of her children. The food she had provided for it is broken and crumbled, and two or three maggots are moving actively and greedily among the fragments. They are not her own. Her own larva is weak and half-starved. It moves feebly and sluggishly and tries without success to win its share of the food. Soon it has no strength left for the struggle and it dies.

One might imagine that the bee would surely observe that these events are scarcely normal. Her nurseries should not be entertaining these alien inhabitants; but she takes no action to remove or destroy the invaders. And when the time comes to seal up the cell, she does so with the same meticulous care which she bestows on other cells duly occupied by the lawful bee larva.

This last attention, however, does not suit the maggots of the fly. It would be highly embarrassing to them to find themselves, when they had completed their metamorphosis into flies, firmly shut up in a cell with inadequate means of digging their way out. Some instinct apparently warns them of this danger and each maggot makes its escape in good time. It is usual, in fact, for the fly larvae to leave the food when the time for pupation arrives. The bee, then, shuts up an empty cell, and the maggots settle themselves in the soil for the long months they must spend as pupae before they emerge the following spring.

This parasite must be deemed highly successful. Fabre has recorded that he dug up all the nests in his largest *Halictus* village—fifty burrows in all. Nowhere in the whole collection could he find one single bee nymph. The whole village was entirely given over to the fly and its progeny. He collected large numbers of these fly pupae and kept them so as to watch their development. There was no change all through the summer and winter; but promptly in spring they woke up and hatched out.

The *Halictus* bee was out of doors already and busy with her burrows. Exactly on time the fly was out, too, ready to watch for her opportunity and carry out the same ravages her mother had inflicted before her.

HALF-WAY TO SOCIAL LIFE

THE *Halictus* bees not only build their nests in
colonies, and thus show a certain leaning
towards communal life, but each family, for the
season at least, shares a common home. The mother
lays her cluster of eggs in the spring and, in the
early summer, her first family emerges. They may
number about ten and they are all female.

These sisters appear to show admirable good
feeling. There is no rivalry, no quarrelling, no
fighting for possession of the best quarters. There
is room in the family mansion for all—they have,
after all, the ten cells from which they have
emerged, and, so far as can be observed, they
respect each other's territory. They may decide
to conduct fresh building operations on their own,
but they do not interfere with the work of another.
When they meet in the common corridor and in
the entrance, as meet they must, they exhibit an
unfailing politeness. They wait their turn in
exemplary order to go through the narrow door.
There is no jostling and no sign of impatience—
the second to arrive invariably gives way to the first.

Since these bees are all female and there are no

males with whom they may mate, it might be thought surprising that they are so busy excavating cells and hurrying in with their burdens of pollen and honey. They have no mates, but it is possible that they have children. At all events, eggs are laid and young hatch out. And these young bees are not always male, as is the case with the unmated worker among the ants when she lays eggs. They are male and female. This is a surprising result of virgin birth but not impossible. Parthenogenesis leading to the production of both sexes is not unknown in the insect world.

It has been suggested that the eggs may be laid not by these bees of the second generation, but by the older mother of them all. This may be so in some species—the question has not been finally decided—but whatever may be the truth of this matter, it is undeniable that this bee has now other duties which occupy her time.

The old mother among the *Zebra Halicti*, even if she lays no eggs, contributes her share to the general good. As each bee goes in or out of the entrance, a door is seen to open and then replace itself. This is the mother, who now offers her large and somewhat bald head to serve as a barrier in the entrance. She stands aside promptly and unquestioningly when the rightful occupants of the burrow wish to come or go, but in between times she shows herself a vigilant guardian of the nest. The reckless indulgence with which she viewed the approaches of the parasite fly is gone with her faded youth.

Now every stranger is suspect. Ants, rival bees who wish to make use of the burrow for their own eggs—any insect, in fact, which dares to present itself is immediately attacked and driven away.

Occasionally a pathetic and bedraggled-looking creature peers hesitantly into the doorway and attempts to creep inside. It is a contemporary of the guardian of the door, a mother whose young have been devoured by the maggots of the fly. Her burrow is empty, but it appears that she feels a need for the bustle of a young and active family around her. An urge is implanted in her to fulfil the task of guardian, but she has nothing to guard. She wanders among the busy, preoccupied bees putting her head inside now this hole and now that. Always she is driven away. Sometimes she attempts to strike blow for blow, but she is easily defeated.

These disinherited mothers grow more bedraggled as time goes on. They begin to settle in one place and seem reluctant to move. Their occasional flights become short and feeble. At last they are no more to be seen. 'What has become of them?' asks Fabre. 'The little Grey Lizard had his eye on them: they are easily snapped up.'

This seems a sad history, but it is probably wrong to waste too much pity on these frustrated mothers. The emotions of sorrow and mourning and regret for what might have been are far beyond their compass. Indeed, it can scarcely be maintained that they can form any picture at all of the satisfactions of which they have been deprived. As they wander about the colony, vainly seeking to enter the populated burrows of other bees, they are blindly driven by forces they cannot understand to try to find the means of performing those functions which are natural to them.

And yet, when all is said, it is still a matter for regret that mothers should lose their families, that a life's work, however humble, should be frustrated. It is still part of the 'tears of things,' even if the insect feels nothing at all.

Meanwhile, the populated burrows are busy and fruitful, and towards the end of July or August there is a change. For the first time males can be noticed flying fussily about near the burrows of some of these *Halictus* bees. They appear to take no interest in the unmated females of the previous generation who occasionally come out to fly in the sunshine. They ignore them—they certainly make no overtures which could be considered a prelude to mating. On the contrary, they seem more interested in the burrows and their contents, for they are continually making descents into them. Some have even been seen to do a little light work in enlarging the entrance to the burrow so that they may pass up and down more easily. This is a thing almost unheard-of among the males of the *Hymenoptera*—whatever else they may do, they have seldom been observed to work.

The explanation of all this is that the newly hatched females of the third generation seldom leave the burrow and mating takes place not out of doors, as is more usual among bees, but down in the cells. Fabre dug up some burrows of the Cylindrical *Halicti* and found unquestionable evidence that this is so.

As the season turns colder, the males make the most of the few weeks which are all that is left to them of their brief lives. If the sun shines, they fly in the old way above the burrows and seek out those flowers which are still in bloom. By October, however, they have grown far less numerous and by November there is not one to be found. They are dead. They have made no provision against winter, and the cold and the rain have killed them. But they have left their seed behind them, for deep in the burrows the impregnated females of the third generation are awaiting the spring. They wait alone, for the old mother and the virgin mothers of the second generation, if mothers they be, are dead like the males. It will be the task of these young females, each in her own burrow, to build up the temporary family community of the next season.

In the spring they emerge from shelter and generally they dispute by force the right to inherit the parental burrow. The defeated sisters are obliged to go away either to build burrows of their own or to take over a deserted burrow, if they can find one. The victor sweeps and cleans and prepares the cells while the weather is still uncertain, and in May, when the sun is hot, she goes out to find the honey and pollen she needs to prepare the little round pellets of food for her young. And so the cycle unfolds itself once more.

There is another kind of bee which leads a semi-social life, the *Allodape*, and this has been described by Dr. Hans Brauns who studied them in South Africa. These bees generally nest in the hollow stems of plants or occasionally in the soil, and Dr. Brauns divides them into three different groups according to the method by which they provide for their young.

In the most primitive species, the mother bee forms little loaves or packets of 'bee-bread,' arranges them one on top of the other inside the hollow stem and lays an egg on each. The larva, when it emerges, attaches itself to its food packet by special appendages and goes on eating it until it is ready to pupate. Each food packet is the right size to nourish one larva. So far, this differs not at all from the method adopted by several of the solitary bees. There is one important difference, however. Other solitary bees divide off each packet of food with its eggs from the next by building a partition in between. The *Allodape* does not do this. Her packets of food lie one on top of the other without any division. The larvae which hatch from the mother's eggs thus inhabit a common dwelling, although she herself never sees them.

The next method of providing for the young shows a further advance in social life. The mother glues her eggs to the wall of the nest, generally near the middle, in a 'half spiral row determined by the curvature of the tubular cavity.' The larvae attach themselves to the wall by their special appendages and turn their heads towards the entrance of the nest. The mother visits the nest at intervals and brings with her a lump of bee-bread which she puts down among the heads. The mother, therefore, remains in contact with her young after they are born and the young feed together on the same portion of food.

In the third group, the mother bee lays her eggs on the bottom of the tube and when they hatch, she brings in bee-bread and feeds the larvae individually. The larva clasps its packet of food which it is able to enjoy all to itself. As the young develop into perfect bees, the daughters stay in the colony and help their mother in her task of feeding the larvae. Eventually the colony contains eggs, larvae and pupae at all stages of development, and since the mother still goes to the bottom of the tube to feed the youngest, the others tend to get mixed up together, although the largest are generally nearest to the top.

It is curious that in this single genus three different stages of development can all be found in existence at the same time. In the last stage, the mother not merely continues contact with her young after they are hatched, but a family group is formed and a big step towards social life has been taken. Brauns suggests that these stages of development may all have derived from the seemingly unimportant omission of dividing walls between the cells in the primitive stage.

The hive bees represent a further development and have evolved a highly organised social life. This is generally thought to constitute an advance on the solitary life of the more primitive bees, and there is little doubt that it is more efficient. A high price may be exacted for efficiency, however, and the life of the hive, though it provides admirably for the welfare of the majority, bears rigidly and harshly on the individual. Puppet of instinct though she may be, possessing little freedom or will of her own, the solitary bee does seem, by our standards, to lead a fuller and more satisfactory life and to be a more attractive creature altogether. It should be added, of course, that it is impossible for a human being even to conjecture the criteria by which an insect might judge of the satisfactions of its life.

ANTS

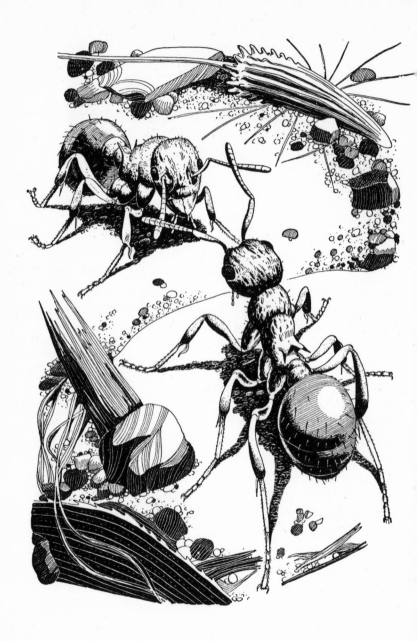

1

WINNING A LIVING

Aɴᴛs in a garden on a summer day can almost
never be observed to pause on a leaf or preen
themselves in the sun. They give the impression
of being perpetually late for an urgent appoint-
ment. So violent indeed is their hurry that they
seem almost to ricochet from one object of interest
to another. There is no difficulty at all in believing
that they are the most agile and tireless runners
in the whole of the insect world.

It may have been the sight of this strenuous
activity which prompted Solomon to urge sluggards
to 'go to the ant.' The Rev. William Gould, an
ant-fancier of the eighteenth century, seems also
to have had a high opinion of their industry. 'The
whole Society is engaged in perpetual incessant
Labours,' he affirms. 'All mutually endeavour to
advance the common emolument, and provide for
the Progeny of their prolific Queen. A colony is
now indeed a small but glorious example of Public
Care. A proper Theme to quicken human Industry
and a just Reproach to the Lazy or Indolent.'
There are moments, however, when one wonders
whether industry may not have become a meaning-

less habit with them, for ants kept in an artificial nest never seem to cease from running about, although this can scarcely be necessary when food and water, which they must procure with so much effort in nature, are provided for them. However that may be, hard work is generally regarded as a virtue and the ants may certainly lay claim to it. On the other hand, ants, however virtuous, cannot be allowed to make their nests in the house or even to become a nuisance in the garden. If they do, it may sometimes be necessary to destroy the nest, and a kettle of boiling water seems to be the accepted method of attack.

As the first tide of water swirls over them and sinks slowly down into the nest, the tempo of orderly bustle changes instantly to frantic haste; but it is not the haste of panic or confusion. Nor is it the haste of those who run away. The ants are hurrying not away from their nest, but back to it. They are returning into danger at the risk of their lives, at the certain cost of pain and burning, so that they may save their brood of young. Soon they come out again, carrying their eggs and their larvae with them, hampered by their burdens but determined, or so it seems, to bear them to safety though they themselves die. With touching courage, they sometimes return again and again to save what they may from their stricken city. At the sight of so much apparent bravery and devotion, it is difficult not to feel convicted of an act of barbarism.

'Bravery and devotion'—these are not words

which may be used of an insect without criticism. They suggest 'anthropomorphism' far too strongly. And yet when an ant goes back into danger to save its young, or even when it pauses to pick them up before escaping to safety itself, its behaviour can be described most economically by saying it behaves 'with courage.' What causes it to behave in this way—whether it is blindly reacting to stimuli, whether it is the slave of certain configurations in its nervous system, certain chemical products of its glands, or whether it is consciously obeying a code of behaviour—this is surely another matter altogether. The same question arises when a man behaves with courage and it has never yet been satisfactorily and finally answered. Some will say he behaves in this way because he has a reasoning soul and wishes to conform to certain ideals of right and wrong which he has formed; others that the pattern of his behaviour is inevitably laid down from birth by the quality and design of his nervous system and his endocrine glands; others that he is chiefly the product of training and environment; others that he is more or less an automaton reacting to the various stimuli which reach him from the outside world.

The difficulty could be evaded both in the case of man and ant by saying that they behave 'as if' with courage, but such pedantry is out of place in a book of this kind. The matter in any case seems of small importance compared to the fact, which is indisputable, that the ant is a social animal which is responsive to social pressures. Like ourselves, it is apparently constrained to do many things which are unpleasant that it may conform to the customs of its community. It is clearly wrong to assume that an ant shares the feelings and emotions of a man, but surely it is equally wrong to assert categorically that it possesses none whatsoever. The fact is that nobody has entered the consciousness of an ant, if an ant can be said to possess consciousness, and nobody therefore is in a position to be dogmatic about its processes.

The standards of self-sacrifice which appear to be demanded of an ant are in some ways rather more stringent than our own, although the cause of her actions may lie in unconscious urges. The starving ant which comes upon a tempting morsel of food will seldom pause to eat it herself, or so it has been said. She will generally bear it back to the nest and share it with her sisters before she indulges herself to the extent of even one mouthful. If the food is liquid, she will fill her crop—her 'social stomach'—and hasten back to her home where she will distribute her prize among her nest-mates, reserving only a very little for her own use. She does this in spite of the fact that the worker ant is a form produced by an inadequate food supply and passes most of her life in a state of comparative hunger.

If a man who was starving found food and hastened back to share it with his companions without waiting to take any of it himself, we should be greatly impressed by his unselfishness and courage. Among ants, such an act is conventional and unremarkable. It would be the exception, in fact, for an ant to do otherwise.

Some authorities say that exceptions might be found mainly among the more primitive kinds of ant—the giant, savage, flesh-eating ants of the tropics—although food-sharing is a social characteristic of all ants. The standards of life and behaviour of these primitive ants are thought to be very similar to those of their remotest ancestors, the ants of Mesozoic times which flourished more than one hundred million years ago, and they have never attained the degree of civilisation which prevails among the higher races of ants. If their life is not particularly short, it certainly tends to be 'poor, nasty and brutish.'

One reason for their low measure of advancement may be the fact that they have remained exclusively carnivorous in their diet. A dependence on the capture of game seems to be incompatible with a high degree of civilisation, perhaps because so much ingenuity and effort must be expended in the pursuit of an elusive and undependable food supply that little energy is left to cultivate the higher social arts. At all events, societies of hunters, whether ant or human, generally remain at a comparatively primitive level, although it may be dangerous to make such a comparison between the two.

The higher races of ants have found other means

of winning their living. Some have taken to a rather specialised form of gardening, some have become harvesters and others have become keepers of cattle and herds.

The 'gardeners' grow fungus in their gardens. It is a fungus of one particular kind which they have succeeded in modifying in a way which suits their needs. They have persuaded it to produce a crop of round, swollen knobs, rather like the vegetable Kohl-rabi—they have been called, in fact, 'Kohl-ràbi heads.' These form the food of the colony, and they have no other.

Originally the ants probably used their own excrement as the basis for these gardens, but little by little it was found that other material was useful as bedding, and some of the fungus-growing insects use debris or wood-shavings. Some species of ant have even learnt to cut leaves from the forest to make their fungus-beds.

One tropical species of such ants divides up the day's work according to each ant's size and capabilities. The largest of the ants leave the nest on organised expeditions to cut leaves for their colony. They ascend the trees in great columns and cut down surprisingly large sections of the leaves they find growing there. When they have cut enough, they return home with their booty, 'their course marked by a line of waving banners, vivid green against the rain-soaked earth,' as one observer has put it.

When the leaves have been brought into the nest,

they are systematically chewed up before being inserted into the gardens. These are laid out on the floors of large underground chambers with arched ceilings, which may be fifty to one hundred centimetres long and about thirty centimetres wide. The ventilation of these underground gardens must clearly present a difficult problem and the ants seem to have been aware of it, for they are careful to provide a number of holes leading up to the surface from these chambers. Temperature is not left to chance either, for the ants close up these holes and then open them again from time to time so as to keep the temperature steady and at a suitable level for the fungus.

Weeding is carried out with great care and attention and, so far as is known, these gardens invariably represent pure cultures of the particular type of fungus grown in them. No alien fungus has ever been found in these gardens. This is the more remarkable since underground conditions would favour the growth of many kinds of fungus and the ants must constantly bring the spores of alien growths into the nest on their rough, hairy bodies. Yet none are permitted to grow.

Other ants have learnt to harvest seeds. These go out and gather up seeds from the ground or pluck them from the plants. Then they bring them home, extract them from their envelopes, throw the chaff and discarded coverings on to the kitchen midden, which they keep outside the nest, and carry the seeds into their granaries. The granaries are generally flat, round chambers which may be as much as fifteen centimetres in diameter and about 1·5 centimetres high. The ants appear to realise that damp will make the seeds germinate and after heavy showers of rain they may carry the seeds to the surface and spread them out in the sun to dry. The 'soldiers' of the colony, when not occupied with more martial duties, use their powerful jaws as seed-crushers.

In some species of harvesting ants, the workers go out to forage individually, but they seem to follow the same tracks, which develop into easily distinguishable roads running for a hundred feet and more from the nest. Obstacles are carefully cleared away from these paths and the vegetation bitten down so that they form regular highways.

Other species of ants have adopted a pastoral life and tend their herds with something of the care and consideration which would be expected of a human farmer. The cattle are generally aphids, or greenfly, but ants will also adopt other insects such as mealy-bugs or caterpillars. All these insects have the attraction for ants that they excrete a liquid which they find pleasant to drink.

Ants have often been watched milking their 'ant-cows.' They crawl over the greenfly apparently without disturbing them in any way, and then stop by one of them and stroke it gently on the abdomen with their antennae. After a while, the greenfly discharges a drop of liquid which the ant licks up with every appearance of enthusiasm.

Then the ant goes on to another greenfly and repeats the same stroking process with the same success. Occasionally the greenfly becomes exhausted by too much milking and the ant gets nothing for her trouble. But she is not defeated. She waits patiently while the obliging greenfly replenishes its supplies by sucking up some fresh sap from the stem on which it is pasturing. The sap consists largely of water and sugar in various forms and is only slightly changed by its passage through the greenfly, so that the liquid discharged is pleasantly sweet. The greenfly are not at all grudging in yielding supplies of their 'milk' and so long as they have anything left to give, they respond to the ants' solicitations at once.

It is impossible to say whether the ants feel any affection for these obliging creatures, but they undoubtedly take trouble to ensure their safety and well-being. They may house them in little sheds or pavilions which they construct for them with great care from earth or silk so as to protect them from the weather and to prevent them from straying. Sometimes they erect a shelter over them actually on the branch on which they are established. When this happens, it has been maintained that they may collect greenfly from neighbouring plants and deposit them all together in a herd, though there is no proof of this. Sometimes, too, they carry the 'ant-cows' to their own nests and keep them there. When they do this, however, they are careful to scrape away the soil from the roots on which the greenfly like to pasture and then they place them on these to feed.

The ants stand guard over their herds and protect them with considerable spirit. If any insect tries to molest their cows, they advance fiercely to the attack with menacing open jaws. Some kinds of ants are able to discharge volleys of formic acid on the invaders, and this they do. If all else fails, however, and flight seems to be the only course left open to them, they seize their charges in their jaws and scurry away with them to a place of safety.

The greenfly, on their side, seem to appreciate the care they receive and make no attempt to escape from the ants which keep them. Some species have even learnt to discharge their droplets of 'honeydew' gradually, in a way convenient to their attendants, instead of throwing them off some distance away with a sudden jerk, as they do when they are alone. The association, in fact, seems to be productive of good on both sides, for the ants also introduce an element of hygiene into the life of the greenfly by removing their excreta in which, left to themselves, the greenfly may sometimes drown.

Ants also domesticate the 'Honey Caterpillar' which emits droplets of a colourless sweet liquid. The ants drink up this liquid, apparently with great enjoyment, and in return they give the caterpillars protection from their natural enemies. One kind of caterpillar—the 'Large Blue'—even passes the

winter in the ants' nest. Naturalists have been
puzzled about the fate of this larva between the
time in August, when it habitually leaves the
plant on which it feeds, and its reappearance in the
spring as an 'imago.' The mystery has now been
solved and the English naturalist Donisthorpe, who
contributed to its solution, has described what
happens. Apparently this caterpillar wanders about
until it is met by an ant which milks it by caressing
its hind end with its antennae. When milking is
completed, the ant walks round the larva and
eventually some signal is given either by the ant
or the larva—it is apparently not quite clear which
takes the initiative. At all events, when the signal
is given, the larva hunches itself up into a shape
convenient for carrying and the ant picks it up and
removes it to the nest.

The larva is scarcely a considerate guest, for it
makes at once for the chambers where the ants'
brood is most plentiful and settles there. Its intentions, however, are not kindly. On the contrary,
the larva eats up the ants' young and grows fine
and large on the nourishing diet they provide.
Soon it spins a cocoon, pupates in the galleries of
the ants' nest and eventually emerges into the open
at the end of June or the beginning of July. The
importation of this caterpillar appears to be a bad
bargain for the ants. The loss of a considerable
part of their brood is surely a high price to pay for a
few drops of sweet liquid; but they are curiously
reckless in such matters.

2

THE LIFE OF THE NEST

THE feelings of an ant, if she may be said to have any, remain impenetrably mysterious. It is difficult to formulate any clear idea of what one ant may feel for another and almost impossible to produce evidence in support of it. There is no doubt, however, that ants recognise their nest-mates and they generally tap antennae in what we may presume to be friendly greeting when they meet. If a strange ant is put into the nest, however, it is threatened at once with open jaws.

There is some evidence that recognition between ants depends largely on smell, for ants which have been bathed with the blood of a strange species, or even with alcohol, are no longer known to their sisters. In time, however, their own characteristic smell reasserts itself and they are recognised and received once more. This smell seems to be manufactured inside the ant in some way and is not picked up in the nest, for a young ant taken from the nest and kept away from it for some time is accepted without hesitation as soon as she is put back into it.

Communication between ants may not include conversation—though we are in no position to state that it does not—but certainly they can pass on news and convey their wishes to one another. Assistance will be rushed from the nest to comrades who have been attacked. An ant who has found a morsel of food too large for her to carry alone will go back to her nest, make clear her difficulties in some way to her comrades and return with a team of workers to help her. The means of communication appear to be gestures of one kind or another—tapping or stroking with the antennae, laying hold of each other with the mandibles, complicated salutations—and also the sound known as stridulation. The ants produce this high-pitched note by scraping one section of their abdomen over another which has a surface covered with fine ridges something like a file. This sound is generally inaudible to the human ear, but with some of the larger species it can be heard very faintly.

Direct action may replace subtle attempts at communication, according to some entomologists, and an ant who wishes another to accompany her to a certain spot may simply pick up her fellow-ant and carry her. The other ant seems to understand what is required and co-operates by rolling herself up so as to make as convenient a bundle as possible. Other entomologists, however, question the accuracy of this observation.

These acts of co-operation seem to show a practical understanding between ants in carrying out the duties of the nest, but there is evidence of something warmer in their relationship. They have a 'crop' behind the mouth-parts which they can use as a storing place for food to be brought up later on and shared with their fellows. They seem to enjoy performing this act of generosity and it is certainly appreciated by the ant who receives the food. She in her turn takes a little more than she needs and distributes some to other workers or perhaps to the larvae in the nest. This constant giving and receiving of food must surely make for good relations and social solidarity and it is a prominent feature of ant life. The signals by which a hungry ant solicits food from one which carries a store, generally by stroking her gently on the head with an antenna, have become well-established among most of the higher races of ants.

Ants have an almost fanatical regard for cleanliness and spend a good deal of their time in the nest scratching away at their bodies with their strigils— the comb-edged spur on their front legs—and licking themselves with their tongues until they have removed every particle of dirt. This service they also perform for one another. Nor are they careless of the hygiene of the nest, for they never excrete in the chamber where the young are kept, but do it far away from them, generally in one chamber apparently set aside for this purpose.

Among primitive ants, the sick and the dying drag themselves away to lonely and deserted galleries of the nest where they may end their lives alone, for only so can they die in peace. Instinct seems to warn the dying ant that her savage nest-mates would fall on her and tear her to pieces for a meal long before life had gone from her, if she sought to die in their company. Among the higher races of ants, it is different. There is little evidence of any care or compassion for the sick or the aged, but at least there is tolerance. If a worker is injured, her injury will be ignored so long as normal activity is maintained. If her injury is serious and she can no longer work, her companions begin to avoid her company. If her injury is mortal, she is picked up and thrown on to the kitchen midden. But she will not be attacked or molested, and, when death comes, her body will seldom be devoured.

This may seem little enough, but it is an advance. And there are intimations of something more. When the young workers go out from the nest to forage for the first time, they may sometimes get lost or they may fall exhausted by the way. It has been said that, when this befalls them, their older and more experienced comrades may pick them up and bear them home on their shoulders.

The care of wounded companions is much more exceptional and some experts doubt whether it ever takes place, but it has been described by reliable observers, and the American naturalist Wheeler has noted the rescue of fellow-ants in distress. He kept a large colony of ants in an artificial nest surrounded by a moat filled with water. From time to time ants would fall into the water and on a number of occasions he saw other ants reach down and pull them out.

In the rare moments when they relax from labour ants may play together. They have been seen to chase each other and pretend to fight. Some entomologists even declare that they do exercises and give each other rides, but others strongly deny that such things have ever been seen or ever will be. However that may be, a colony of ants appears to be a peaceable, united and happy community, but it is by no means a community of equals. At the head of all, the queen, though in fact nothing but an egg-laying machine, has her supreme position, apart and unique. The lazy and unco-operative males are tolerated in the nest for no other reason, one suspects, than that the ants, so resourceful in all else, have so far found no means of doing without them. The third group, the workers, are often very varied both in size and function and they range from the large and formidable 'soldiers' to tiny minims, scarcely one-third of their size. These workers are female in sex, but not noticeably so in function, although some of them may lay eggs towards the end of their lives. Their eggs, however, are self-fertilised and as a rule produce only male progeny. Gould esteems this inability to lead a normal sex life a 'particular instance of divine wisdom. Had the common workers been of either Sex,' he remarks, 'it must have given a great deal of Interruption to their Labours.'

For most of their active lives, the efforts of the workers are directed towards the maintenance and well-being of the existing community, and the responsibility of reproduction is left to the queen alone. Each worker has her special duty, though it may not be the same for life. The youngest workers, the callows, generally undertake the task of caring for the young. This is a sufficiently strenuous task for, apart from the obvious duty of feeding and cleaning the brood, it seems to be thought necessary to move them about the nest constantly, almost from hour to hour, and arrange them first in one apartment, then another. The different age groups are not mixed up together but are kept severely separate, apparently because their temperature requirements are thought to differ. The pupae in particular are considered to need more heat than the young in other stages of development and some ants even build incubators of vegetable detritus to bring them on.

The British entomologist Donisthorpe has described the way in which some ants he kept in an artificial plaster nest would pile up their pupae in the chamber nearest the fire. There they would watch over them carefully, apparently noting their progress, for from time to time, he says, a worker would 'suddenly seize a pupa and hurry off with it, as fast as she could, back to the cooler chambers, as if it was a joint before the fire, and was just cooked to a turn.'

These duties, then, may be onerous but they do not take the young callow ant from the home and she is not exposed to all the hazards and perils which attend expeditions abroad. When she has once passed this stage, however, she may find herself committed to different tasks, according to her size and capabilities. The choice of a job, in fact, appears to be decided more by physiology than by whim or personal taste on the part of the individual ant.

The 'soldiers' are obviously destined for their role in life by their size and their powerful jaws. These larger ants may also use their jaws to perform the tasks of peace, as they do among the harvesting ants, and in one Texan species of ant they sacrifice their large stopper-shaped heads to provide a movable front-door to the nest. Their heads exactly fit the circular hole which forms the entrance to the nests of these ants and one of these soldiers is always on duty with her head pressed into it. When a worker wants to go out, she strokes the soldier from the back, who then stands aside to let her pass. When she returns she strokes the soldier's forehead, and again this animated door obligingly moves out of the way and allows her to enter. It would seem a monotonous occupation, if undemanding, but one for which the individual ant would be inevitably marked out by the peculiarities of her physique.

The curious vocation adopted by some of the honey ants is more puzzling, for here some kind of individual choice does seem to play a part. A group of young callow ants of this species all seem much alike. When kept in captivity, they feed greedily on maple syrup and cane sugar water, but it is difficult to show that any one of them takes more than another. Yet out of any random group of young workers, one or two will begin to assume the appearance and function of 'repletes.' Gradually they lose the slender figure of the typical ant and they grow rounder. The abdomen swells and swells until at last it is distended like a balloon, and the ant is scarcely able to move at all. Helped by her companions, she climbs up to the ceiling of a specially prepared chamber and suspends herself by her claws. There they hang in rows, these curious 'repletes,' filled to capacity with honey— so tightly filled, in fact, that the walls of their stomachs may be ruptured almost by a touch. And so they devote their lives to acting as a living storage tank for the benefit of their fellows. When a worker needs food, she will go to this unusual store-room, stroke a 'replete' with her antennae, and receive an offering of honey. In return, the workers feed the 'repletes' and if any of them should be so unfortunate as to fall from the ceiling and miraculously survive the crash without bursting, the workers strive to hoist the almost helpless creature back to her perch.

The presence of 'repletes' in the nest is obviously a convenience, since they provide a reservoir of food against hard times and shortages. This is understandable. It is far more difficult to understand the inner force or urge which impels certain ants out of any one brood to feel that they must fulfil this unique destiny, while their sisters grow up to become active foragers and workers. Does the difference lie in the structure of their abdominal walls, as some experts have suggested? Is the difference in their glands or in their genes? Can it be said that their choice of vocation is psychological, and that some ants, by their temperament and character, incline naturally to a life of inactivity and contemplation? It is as idle to ask this as to seek to know exactly why one man chooses to be a test-pilot while another prefers to spend his working days reading in a library.

These somewhat exotic specialisations are the exception among ants. For the average mature worker, the choice, if choice there be, lies in the main between the ardours of foraging and the hard labour of building and maintaining the nest.

The architecture of an ants' nest, though it may be untidy, is perhaps the more interesting in that ants do not cling to a slavish rigidity of design in the way that a bee or wasp is inclined to do. Ants will take advantage of the natural contours of their chosen situation and the finished nest often appears to have been designed in such a way as to make the most of existing features. There is even a certain whimsicality in the design of some nests which has been conceived by some experts to reflect personal taste, though this is dismissed by others as pure imagination. At all events, the finished structure may be ambitious to a degree, and Gould found much to praise about it.

The manner of their Architecture deserves our
Consideration (he writes), as being adjusted with remarkable Curiosity and Art. The whole Structure is
divided into a Number and Variety of Cells or Apartments, all communicating with one another by little
subterraneous Channels which are circular and smooth.
The Smoothness is commodious to the Tenderness
of the Young which they frequently carry from one
Lodgment to another.

We cannot less admire the Texture of their Cells.
As the Ants lie together in Clusters and dispose of
their Eggs and many of the Young in the like manner,
an oval Figure is the most convenient for this purpose,
and such is the Structure of many of their Apartments.

There is no doubt that Gould attributed much
to conscious art that is, in fact, due to chance. All
the same, although some of their apartments are
small and unambitious, others are spacious and
imposing with arched ceilings supported either by
little columns, by slender walls or by strongly-
formed buttresses. All this is accomplished without
the aid of tools, for ants have never cared, or have
never been able, to learn their use. It may be that
they manage so remarkably well by using those
implements with which nature has endowed them
that they have never felt any need to supplement
them. They work chiefly with their mandibles—
powerful curved fangs with rough toothed edges
placed at either side of the head—and their claws.
If ants wish to build on to their nests, which are
generally just below the ground, they take advantage of a shower of rain, as this brings the earth to a

suitable consistency for making their pellets of
soil. In dry weather they may be obliged to bring
water from a distance. Ants generally add another
storey to the top of their dwelling, if they want to
make it larger, and they can sometimes be seen
emerging from below the ground, each carrying
pellets of earth between its mandibles. They
mould these pellets with their fore-legs into something not very far removed from a brick, then they
place these 'bricks' into position with their mandibles, and pat them down with their fore-legs.

It soon becomes clear, maintains the French
authority Huber, that they are not each following
some random urge, but that they are working to a
concerted plan. Others, however, would dispute this.
As the walls rise, the shapes of galleries and apartments and of the larger chambers begin to declare
themselves, according to Huber. The ants build up
the walls to a certain height, piling brick on brick,
and they then turn themselves to the task, which
has perplexed the architects of all ages, of bridging
the gap between one wall and another in some way
which will ensure stability and security in the
finished structure.

Fragments of suitably moistened earth are
attached to the tops of the walls and soon it can be
seen that building is now proceeding horizontally.
A ledge is formed on the walls which lie opposite
each other and these ledges are extended until they
meet. The galleries are a bare quarter of an inch in
width, as often as not, and pose the ants no great

problem in construction. They seem equally undismayed, however, by the task of spanning chambers a good two inches or more across. The ceilings of these chambers may be vaulted, and the ants may build pillars to support them. In some cases, they may strengthen their walls with buttresses. At all events, the problem is eventually solved in some way that seems satisfactory and another storey is added to their dwelling place.

The materials of this type of building are plentiful and easy to obtain and in consequence ants are not so tied to their homes as social bees or wasps. There is no great inconvenience involved for them in deciding to make a move, and this sometimes happens. The change generally reflects the wishes of the more enterprising and adventurous workers. These, apparently impressed by the advantages of some new location, enter on the considerable work of transporting the brood, the queen, their fellow workers and indeed all the inhabitants of the nest to the chosen site. These other workers, who as often as not have been picked up bodily and obliged to change their home in this somewhat rude way, begin to pine, or so it would seem, for their old and familiar surroundings and themselves enter on the laborious task of carrying the brood all the way back again. The supporters of change find them at it, snatch them up together with the brood and dump them once more in the new quarters. And so it goes on.

This disagreement as to the most desirable living

place for the colony may persist for days or weeks, and all the time files of ants will be making their way between the two nests, carrying the long-suffering pupae and larvae in their mandibles. At last, however, the wishes of the progressive party prevail, the others either fall in with their ideas or resign themselves to inevitable change, and the colony becomes established in its new home.

In spite of this desire for change on the part of some ants and the comparative ease with which new quarters can be constructed, most ants seem to be attached to their homes. They seem willing to spend considerable time and trouble in caring for their apartments, polishing the walls and making small improvements, and they keep their nests scrupulously clean. Almost every nest has a kitchen midden on to which all the rubbish is thrown, including the dead bodies of their comrades. Pliny maintains that the ants of Italy bury their dead, but this custom, if it ever existed, does not seem to have spread to English ants.

Other kinds of ants build their nests in the cavities of plants or in woody plant tissues. Others again establish themselves in houses or even, rather adventurously, in ships. Some of the tropical ants, however, are really ambitious and construct their homes high above the ground in trees, using either earth, 'paper', or silk. The earthen 'ant-gardens' are generally spherical in shape and look something like a bath sponge suspended in a tree. The 'paper' nests may be enormous and are said to be sometimes 'large enough to enclose the body of a man.' These, too, are generally hung from trees, although they may also be found in logs or under stones.

The silken ant nests are perhaps the most ambitious of all and are generally constructed of leaves. One of the largest types is found quite commonly in tropical Africa and consists of 'a number of leaves fastened together by a fine white web, like the finest silk stuff.' Here, according to Ford, 'the large, long, very vicious, reddish and greenish, worker ants live, with their grass-green females, their black males and the whole brood.' The adult ant is unable to produce silk, but these ants exploit the silk-spinning capacities of their larvae for the construction of the nest.

In the midst of all the activity and bustle of a thriving ants' nest the males alone potter about the galleries with no task to occupy them, lazy, stolid and at ease. The busy and preoccupied workers find time to give them the food and care they need, for they are incapable of providing for themselves, and they carry them to safety, if danger threatens the nest. The workers treat them, in fact, exactly as though they were immature larvae, and they leave them to sleep out their days, if they wish, without attempting to urge them to any useful occupation.

The virgin queens who, like the males, possess wings are hardly less idle. The ants have shown the social wisdom not to depend on rearing one queen alone, and there are generally several growing up at one time in the nest. On the whole, they lead quiet, uneventful lives, but they are not so inactive as the males, for they have been observed to take their turn with light duties in the home. They do little real work, however, for they are conserving their strength for a considerable ordeal.

As the time for the mating flight draws near, the young queens begin to grow restless. The males, for their part, are quite transformed. They throw off their sluggish inertia, they run alertly about the nest and appear to be suffering considerable nervous strain. From time to time the queens and males may surge to the entrance of the nest and try to escape into the air, but they are frustrated in this enterprise by the watchful workers, who apparently judge the attempt to be premature.

When a day comes which the workers deem suitable for the mating flight—or when climatic conditions provide the necessary stimulus—they make preparations for this great event in their lives. Sometimes they excavate special galleries leading to the open air, and even the blind workers belonging to those species which pass all their lives underground come hurrying into the sunshine on this one occasion alone. With triumph and joy, as it seems, the great multitudes of the workers accompany the young queens and their mates to the surface. Here they press round them and seem to be treating them with deference and attention, grooming them for the ordeal of their first and only flight. Other workers climb to the tops of blades of grass, as though to make sure of an advantageous post from which to watch the spectacle to come. Since they have very poor eyesight, however, it is doubtful whether they see it. Meanwhile, the male and female ants, the only ants to possess wings, seem to have become almost tremulous with excitement and they run here and there in indecision, seeking to launch themselves upwards, tentatively spreading their wings.

The males are generally first in the air, and these sluggish creatures of the nest now show an extraordinary energy, as they beat upwards in search of their mates. The queens follow them more slowly and rise gradually higher as they use their wings for the first time. They climb far up into the blue sky for the great moment of mating.

The workers are left behind on the ground. They seem to be waiting until there is no more to be seen and they can no longer follow the progress of the winged fliers, then they turn slowly back to the nest, seal up the special galleries they have opened and go about their unpretentious tasks. It is no doubt fanciful to suppose that they must feel a little flat, a little wistful. They have had their small share in one of the supreme moments of their colony, they have helped forward an experience which they can never themselves know. Now they must turn to other matters. For them, the affair is over and done with, for it is unlikely that they will ever see their queens or their males again. The destiny which they must fulfil is no longer a concern of the nest.

The male ant, unlike the male of the honey bee, suffers no violence at the time of mating. His male organs are not torn from him, he returns to earth whole and unharmed, but his remaining days of life are none the less few. He cannot find his way back to the nest, so he creeps under a stone or hides himself under a twig so as to procure a shelter for the night. When day comes, he may fly again for a short time, he may even succeed in finding a little pollen to eat; but his strength soon fails. He has never been used to work for his bread, he has never learnt the laborious business of providing for his daily needs, he is ill-equipped for life. He has performed the one function for which he was born, and now, very soon, he dies.

The destiny of the queens is different. They return to earth to face an ordeal which only those who are very strong and very capable, and perhaps a little fortunate as well, may hope to survive. In the air they have mated and they have received sperm which will last them all their days. They will never mate again and they will never fly again. They are marked out to be the founders of a race. But that destiny lies on the other side of the ordeal with which they must now contend.

The queen floats down to earth and rests for a little in the sunshine, quietly polishing her bright armour and gathering her strength. Physical pro-

cesses are at work within her which will show her
what she must do next.

She pulls off her wings. It is a symbolical action,
but practical, too, for from her degenerating wing
muscles will come the nourishment for her first
brood. She struggles free of her wings, pulling at
them with her legs and rubbing them against grass
blades and pebbles. At last they flutter to the
ground and she takes up her life on the earth which
she will now never leave.

She has much to do. First and most urgently she
must find shelter and a convenient hiding-place
for the work before her, and for this she rapidly
examines rock crevices, looks under pebbles, searches
in holes. At last the site is selected and she begins
to excavate her burrow. Before the first night is
over, she has probably constructed her cell and
blocked up the entrance; but she has paid a price
for this achievement. The teeth on her mandibles
may be worn away, the hairs rubbed from her body
and her sculptured, shining armour battered and
dull. But she now has a home.

From this time she will be completely alone for
many weeks, or even months. Drought may kill her
or cold or flooding. She may be attacked by other
creatures which live in the soil, or fungus may invade
her body and destroy her. Only the very strongest
of the queens of these species survive this dangerous
period of solitude which must be endured while she
waits for her eggs to mature inside her. She has
no work to occupy her, but she may make small

improvements to her burrow, or polish the walls. She sees no living creature and she has no contact with the world. She eats no food.

At last she begins to lay her eggs. Still alone, she tends them and cares for them. The larvae hatch out of them, and these she feeds with her saliva. In the nest, the queen was fed richly and she stored her food in the form of fat and bulky wing muscles which filled her whole chest. By the time her first brood is reared, nothing will remain of them but an empty cavity, a testimony to the sacrifice she has made. For it is the fat which comes from these decaying wing muscles which enriches her saliva and nourishes her young. It is a meagre diet, but it suffices.

The larvae grow slowly on this food and their transformation into pupae takes place sooner than is normal. As these tiny pupae come to maturity, the mother, still alone, still unaided, still unfed, helps them from their wrappings, and the first brood of workers is born.

They are small, the product of near-starvation. To feed them at all, the mother has sacrificed her own flesh. But her reward is to come. The tiny workers are active and restless and normal in everything but their size, and soon they dig their way out to the sunlight. They come back with the food they have won, the spoils of their first day's hunting. And for the first time in many months, it may be, the queen-mother is fed.

Her supreme solitary ordeal is over. Now, little

by little, life will become easier for her. Soon she will take no interest in her brood, show them no care—all will be done by the workers. She will be fed and cosseted and attended, her one task the laying of eggs which will continue her whole life long. Her pioneering days are done. She has founded a new colony.

As the colony begins to thrive, the queen seems to change. She becomes timid and retiring, shrinking away from the light or from any disturbance. She degenerates into a mere egg-laying machine so that it becomes difficult to remember that she once faced such dangers and survived such strenuous endeavours. It would be pleasant to imagine that her progeny remember her triumph and appreciate, after their fashion, the extra-ordinary sacrifices made by their queen-mother in order to found their city and launch their community on the world. It is highly improbable, not to say impossible, that ants are capable of any such recollections, but none the less the queen is treated with great attention by her workers. Indeed, she could be forgiven for feeling the attention to be somewhat exaggerated. She always seems to be the centre of a milling crowd of ants which press themselves against her and stroke her ceaselessly with their antennae. If she wishes to move, she must push herself against the mass of insects which never seems to part to open a path for her, but at the most moves slowly backwards in front of her. It may be what she likes, but one imagines it might become a little trying. Gould, as always, takes the romantic view of these proceedings.

> In whatever Apartment a Queen Ant condescends to be present (he writes), she commands Obedience and Respect. An universal Gladness spreads itself through the whole Cell, which is expressed by particular Acts of Joy and Exultation. They have a peculiar Way of skipping, leaping and standing upon their Hind Legs and prancing with the others. These Frolics they make use of both to congratulate each other when they meet, and to show their Regard for the Queen. Some of them gently walk over her, others dance round her, and all endeavour to exert their Loyalty and Affection.

Even when she dies, it seems, her body is not thrown upon the kitchen midden like that of a common worker, but care and attention will be offered it so long as it is recognisable at all.

The queen of the fungus-growing ants has a task to complete after her mating flight which is more severe still, for she must found a garden as well as a family. When she leaves the nest, she retains in a special pocket behind her mouth-parts the remains of the last meal she has eaten. As soon as she has finished digging her burrow after the mating flight, she lays this pellet on the earthen floor and so sets in train the cultivation of a new fungus-garden. She cannot go out to get leaves or debris from which to form a bed, so she sacrifices some of her first eggs to her garden. She breaks them up and scatters them as manure over the slowly-growing fungus. Sometimes she pulls up a little piece of fungus with her mandibles and presses it against her body to receive a tiny yellow-brown droplet which she excretes on to it. After she has enriched it in this way, she plants it again and pats it carefully in place with her feet. So, in time, the garden begins to flourish and is ready to support the new colony.

These are the heroic queens. The queens of some other species do not find it necessary to immure themselves in their burrows until the first brood of workers is mature, but go out to forage for food instead. Queens of other species again are poorly endowed and seem to feel unequal to the great task of founding a new colony without assistance. At all events, they do not attempt it, and strive to associate themselves, when mating is finished, with adult workers, preferably of their own species. When this happens, they simply increase the progeny of the nest and lengthen the life of the colony.

Sometimes, however, the queen comes down from the mating flight on ground which does not hold any nest of her own species, and she is obliged to try to make her way among ants of an alien kind. She may be lucky enough to find a queenless colony which will adopt her, feed her and rear her young. The queens of other species enter an established colony and murder the queen. Some attack the alien workers, massacre them and then steal their young, so providing themselves with a ready-made family which will bring up their broods for them. The queen of one species follows a more cautious plan. She climbs on top of the resident queen where she is safe from attack, waits until the workers have accepted her presence and then saws off the head of her rival.

The queen is distinguished from ordinary worker ants not only by her unique function in the nest, but by her size and her length of life. A queen ant may live for as much as fifteen years, but a worker survives for no more than seven at the most. In spite of their differences in adult life, however, queen and workers alike begin as eggs which are indistinguishable. No special size or conformation of the egg marks out the future queen. The eggs hatch into larvae, and here again, to the human eye at least, queens and workers seem all alike.

It was thought at one time that the feeding of the larvae might determine whether they developed eventually into queens or workers, but although there is some evidence that workers may develop from underfed larvae and may be, to some extent, 'hunger forms,' there is nothing to show what causes differentiation between the workers themselves. Nor can it be shown that well-fed larvae must develop into queens.

In many species, the feeding of the larvae is far too haphazard to suggest that any predetermination is exercised by the ants themselves. Among the more primitive ants, the larvae are fed on insects which are chopped up and scattered among

them. 'This evening,' writes Wheeler of a nest of
Ponerine ants, 'several horseflies were at once
shorn of their legs, then decapitated, and finally
their thoraces and abdomens cut into smaller pieces
and distributed among the larvae. One was given
a fly's head which it kept twirling round in a
comical manner . . . another was given a piece
of thorax, still another a leg with a mass of muscle.'
He describes the way in which the worker ants
carried crumbs of cake to the larvae and allowed
them to feed. Their meals were frequently interrupted because other workers would keep hurrying
up and snatching the larvae away. When this
happened, the cake would be at once picked up by
other workers and offered to other larvae. In a short
time, the cake had circulated round a very large
number of larvae which had all taken more or less
snatched meals from it. But while these larvae were
thus able to enjoy only rather intermittent and
unsatisfactory periods of feeding, one small larva,
which was lucky enough to escape the attentions
of the workers, fed steadily and uninterruptedly on
a morsel of housefly.

Such observations certainly do not suggest any
concerted plan of feeding designed to rear queens
on the one hand or workers on the other. The problem, in fact, of how the different forms arise is still
largely unsolved.

The larvae of other species of ants are fed chiefly
on the food which is natural to the species. The
harvesting ants present their larvae with fragments

of seeds and these larvae possess specially developed mandibles to deal with this tough resistant food. The fungus ants feed their larvae on wisps of fungus and the ants of practically every species give them their share of regurgitated food.

Whatever the food offered, the feeding of the young is seldom neglected by ants, but they cannot be wholeheartedly praised for this, for there is little doubt that self-interest plays a part. The larvae exude saliva and most of them also provide a fatty secretion which comes out either through their skin, or, in certain species, from special appendages. These substances are apparently most attractive to the workers who lick the larvae with great persistence in the hope of obtaining supplies. This incidentally benefits the larvae as it frees them from moulds which might otherwise be harmful.

Among the primitive ants, the workers behave with a certain brutality to the larvae in their determination to get a supply of these secretions, and they have been seen to pinch and ill-treat them for hours at a time, without giving them any food at all. The more advanced kinds of ants seem to have learnt that the larvae exude these substances more freely if they are well-fed, and that the best time to get a good supply from them is immediately after feeding. As a result, a fair and just working arrangement seems to have developed by which the workers first feed the larvae and then tap them for a share of these highly-prized substances in return. Thus both parties profit, and the larvae appear to be treated with care and affection by their nurses, however much greed and self-interest may be operating in the background.

Unfortunately, however, hunger and affection are closely linked emotions in all animals, and ants upon occasion have been known to devour their larvae and their pupae. Eggs are eaten even more frequently and it may be their importance as a possible food supply which induces ants to lick them and fondle them, to arrange and rearrange them over and over again with such seeming tenderness. It may indeed be this connection between the young and the food supply which induces ants to make such sacrifices in order to rescue them when the nest is in danger. It is noticeable that they often rescue morsels of food along with their brood of young.

The ant larva is a soft, legless grub, more or less cylindrical in form and consisting of a head and thirteen segments. It appears to be without eyes, but possesses distinct mouthparts, including a pair of mandibles. Sometimes the larva is naked, but more frequently it is covered with hairs which may be long and rigid, flexible and tipped with double hooks, or sometimes forked. The hairs, in fact, vary considerably in the different species and it is thought that they may serve a useful function in protecting the larvae from being chewed up by their fierce and hungry sisters. In some cases, too, the hairs are useful in holding the larvae away from direct contact with the damp soil, or in anchoring them firmly to the walls of the nest. They also serve to hold the larvae together in clumps so that they can be carried about more conveniently by the workers in their constant transportations from one chamber of the nest to another.

When the larva of certain kinds of ants approaches the stage of transformation into a pupa, it spins a cocoon. The workers bury it in the earth at this stage, since a larva clearly cannot spin an envelope about itself if it is lying in the open. It must have walls close about it to which it may attach its threads of silk. When the larva has lined the cavity in which it is lying with a silken web, the workers dig it up again, remove carefully every particle of soil which may be clinging to the outside of the envelope and leave it to pupate.

In one tropical species of ant—*Oecophylla smaragdina*—the larvae can spin, but they do not spin cocoons. They are obliged by their older sisters to provide sewing materials for the nest. These ants build their nests among leaves by fastening the leaves together with silk (the process by which this is carried out has already been described). A naturalist travelling in the district where they are found opened up the nest of a colony of these ants and noticed that, while the majority seemed to be preparing to defend their home against him, a small party went off to repair the tear he had made in the wall. A row of them lined themselves up on one side of the tear and with their claws pulled the other side towards them until the two sides were touching.

Another party then came along and cleared away the torn fragments by biting them through with their mandibles and then tugging them free. The shreds of debris were carried to an exposed part of the nest and allowed to float away on the wind.

The next stage was marked by the arrival of a party of workers from the interior of the nest, each carrying a larva in its mandibles. They walked up to the opening, appeared to exercise some pressure on the larvae by squeezing their bodies in the middle, and then, pointing the foreparts of their charges forward, they moved them from side to side of the rent. As the larva touched each side, a worker seemed to press something down, as though it were attaching the sticky thread spun by the larva to the leaf, and after a little it could be seen that the tear was indeed being filled in with a web of finely spun silk. 'There could be no doubt,' this observer comments, 'that the ants were actually using their larvae both as spools and shuttles.'

It seems that these larvae become so exhausted by being used in constant building operations that when the time comes for them to pupate, all their silk is gone. They seem obliged, therefore, to pupate unprotected by a cocoon, although they belong to a species among which the spinning of silk is a natural activity, and a cocoon might be expected.

Among other kinds of ants, the larvae are not accustomed to spin a cocoon and for them the process of transformation into a pupa is in consequence much simpler.

After a period of time which seems to vary a good deal, the pupal stage is completed and the 'callow,' the young ant, hatches out. Among the more primitive ants, this is generally accomplished without any outside assistance. The primitive *Ponerine* ants, in fact, show no interest at all in their pupae. Their interest is aroused, however, as the young ant emerges, for fragments of the pupal skin may be sticking to her and these provide a meal. In consequence, she is soon stripped of them, and this is much to her benefit.

The higher ants show a more constant solicitude for their young and give even the pupae a good deal of attention. When the time for hatching-out comes, the pupae are helped to free themselves of the enveloping cocoon and pupal skin, and this has resulted in their becoming more helpless than the young of primitive ants and requiring assistance in carrying out tasks which the more primitive ants are able to perform for themselves.

The period of 'growing up' generally lasts for about another year in most species, during which the callow ant leads a comparatively sheltered life, and then she is regarded as fully mature and able to play her part in supporting the life and well-being of the community.

3

SLAVE-RAIDING AND WAR

Not all ants live at peace with their neighbours and some of them are in the habit of carrying out regular slave-making expeditions against other species of ants in order to capture their brood and bring it home with them. Prominent among these are the romantically-named 'Blood-red Slave-Makers' which are unanimously regarded, according to Wheeler, as 'one of the most gifted and versatile of ants.' This ant also has an undeniably ferocious streak, for it attacks an intruder not only by biting it with its mandibles but by injecting formic acid into the wound as well.

Their raids are chiefly directed against the peaceable and less warlike Black Ants. Scouts are sent out into the district around the nest and they search for suitable settlements of Black Ants to attack. By some means, they communicate their discoveries to their sister ants and a decision appears to be arrived at as to which settlement to attack and when. Generally the army sets off to battle in the morning and gets back in the afternoon, but on occasion they may make a much later start than this.

The French entomologist Huber has written a vivid account of a slave-raid.

> On the 15th July (he writes), at ten in the morning, a small division of the Blood-red Ants was dispatched from the garrison and arrived at quick march near a nest of Black Ants, situated twenty paces distant, around which they took their station.

The Black Ants came rushing out at once to attack the strangers and managed to take a few prisoners. The Blood-red Ants, however, made no effort at retaliation but appeared to be waiting for reinforcements to arrive. Small contingents joined them from time to time, and soon they drew nearer to the stronghold of the Black Ants and seemed more prepared to engage them in battle, although they still showed no apparent eagerness for the fight. The Black Ants drew themselves up in front of their nest to meet the enemy and minor engagements became more frequent. The Blacks were generally the aggressors, but none the less seemed to lack confidence in their ultimate victory.

> They look to the safety of the little ones confided to their care (writes Huber), and in this respect show us one of the most singular traits of prudence of which the history of insects can furnish an example. Even long before success is in any way dubious, they bring the pupae from the subterranean chambers and heap them up on the side of their nest opposed to that where the Blood-red Army is stationed, in order to carry them off with the greater readiness should the

fate of arms be against them. Their young females escape on the same side.

The Blood-red Ants, evidently feeling that their reinforcements were now sufficient, launched an attack on the Blacks, and reached the gates of the city. For a few seconds the issue seemed to be in doubt, but then the Blacks apparently lost heart, seized their pupae and made off with them. The Blood-red Ants pursued them vigorously and tried to snatch the pupae away.

The whole body of the Black Ants was soon in flight, but a gallant few fought their way back into the city and tried to rescue the larvae still remaining inside which would otherwise have been carried off as loot. The Blood-red Ants had by this time penetrated right into the centre of the city and established themselves there.

Reinforcements of Blood-red Ants were still arriving and these began to carry away all the larvae and pupae which remained, 'establishing (as Huber says) an uninterrupted chain from one ant-hill to the other: thus the day passes and night comes on before they have transported all their booty.' Some of the Blood-red Ants were left behind in the city when darkness came, and on the following morning they set to work again on the job of removing to their own stronghold all the pupae and larvae which were left. After that, they carried each other home.

Slave-raids have only been described a few times by English writers, but one of these was greatly impressed by the fierce aspect of the soldiers waiting to attack. They constantly assumed the 'most threatening attitudes,' he says, 'occasionally spring-ing up on their hinderfeet and snapping their jaws with great ferocity.'

The advance on the nest to be attacked is gener-ally made in fairly loose and straggling formation, but the ants none the less follow a remarkably direct route to their objective. This is the more striking in that they are not always led by the same ants. Those which have been in front drop back from time to time and others move forward to take their places so that the knowledge of exactly where they are going seems to be somehow diffused over a great many of the ants. It seems probable that scouts must reconnoitre the territory for some time before the expedition and that they are suffi-ciently numerous to control the direction taken by the party as a whole.

The hesitation before going in to attack which Huber describes seems to be typical of the Blood-red Slave-Makers. They nearly always seem to wait some time in front of the nest they intend to attack and send for batch after batch of reinforce-ments before nerving themselves to descend into the stronghold of the Black Ants. The Black Ants, on their side, are generally defeated without much difficulty, although they make great efforts to break through the cordon of the attackers with their larvae and pupae in their mandibles and run des-perately to the tops of blades of grass in their efforts to save their brood. When they are seized by the enemy, however, their resolution quickly fails and they drop their charges. The battles, in fact, are singularly bloodless and the attacking ants, although they are kidnappers by habit and occupation, seem to carry on their trade in a spirit of kindliness. They seize the escaping Black Ants in their powerful mandibles, but so long as these let go instantly of the pupae they are carrying, the mandibles are not brought together and the adult ant is allowed to go free. It is only if resistance is made that the mandibles pierce the head or thorax of the helpless Black Ant, and murder is done. Very often a tremendous encounter appears to be taking place, a desperate battle to be waged, but at the end of it all there is not a single corpse to mark the place.

The marauders strip the captured settlement of its larvae and pupae and proceed to carry them home. Their demeanour appears to express the utmost satisfaction with the day's business and they proceed on the homeward journey with a 'peculiar cantering motion.' When they arrive at the entrance to the nest, they are greeted by their resident slaves, who appear to show an excited interest in the booty and to congratulate the warriors on their prowess. The captured pupae and larvae are then carried inside and a fair proportion of them are eaten—it is probable that hunger may be one of the impelling motives for the raid. The rest, however, are carefully tended and nurtured and grow up to work as slaves in the colony of their captors.

Meanwhile, the ravaged Black Ants are left to do what they may in the face of devastation and disaster. They gather together at some spot near their ruined nest, under a heap of dead leaves, perhaps, or a twig. They have their queen. They have carried in the few larvae and pupae they have managed to rescue. The more active ants have picked up the exhausted and the lost and have gathered them in with the others. But it is little enough. All is to be done again, the building of a colony, the rearing of a new generation. And always before them there is the threat of a new visitation, another descent in the heat of the afternoon, the pillaging afresh of their young, so laboriously reared. It must be hoped that entomologists are right in denying to ants the gift of foresight.

Colonies of ants which have been plundered season after season appear to accept their trials with more philosophy than those species which are less accustomed to attack. These may put up a savage resistance and battles have been known to last for hours, and even for days, with much bloodshed and loss of life on both sides.

The Blood-red Ants—*Formica sanguinea*—use their slaves chiefly to look after the growing broods of young. The slaves may also go out to forage occasionally, attend the aphids or 'ant-cows,' if there are any, and they may perhaps do a little excavating. They are physically weaker and less gifted than their captors and cannot perform these tasks so well as an equal number of *sanguinea* workers, but they are none the less useful and release their mistresses for more difficult and important business. Decisions to remove the nest from one site to another, and the actual migrations, seem to be the responsibility of the *sanguinea* ants who generally carry the slaves from the old nest to the new. The nest seems to be constructed in a manner typical of *sanguinea* ants, so that the slaves, if they play any part in the building of it, must work in accordance with the plans of the dominant species. The *sanguinea* ants, in fact, have not been corrupted by the possession of slaves. They have retained their initiative and skill and there is nothing which the slaves can do which they could not do better themselves, if they had a mind to. It is the Black Ants who are changed, for these normally timid creatures sometimes become quite aggressive in imitation of their captors.

Slave-making, however, has had the corrupting effects on other species of ant which must nearly always follow the practice. *Polyergus rufescens*, for instance, the Amazon ant, can no longer exist without her slaves to help her carry out the humble but necessary tasks of daily life. She is a large and handsome insect, rich brownish-red in colour, alert, vigorous and extremely competent at the one job she can still perform. She can capture slaves with far greater brilliance and despatch than her Blood-red sisters.

The *Polyergus rufescens* warriors assemble in readiness for a slave-raid outside the entrance to their nests. They march to their objective rapidly and compactly, unlike the red ants, giving an impression of almost feverish haste. They show no hesitation when they reach the colony to be attacked, but fling themselves straight into it, still holding their compact formation, seize the enemy brood and make for home again with as little delay as possible. If the defenders offer any resistance, they slaughter them at once with none of the hesitations shown by the *sanguinea* ants. As a result, the carnage attending these raids is often considerable.

At home, however, their behaviour is very different. Their mandibles, sickle-shaped, toothless and pointed, so eminently well-suited to slaughter and battle, are useless for the humdrum tasks of digging in the soil or picking up the tender and thin-skinned larvae and moving them about the nest. They cannot even feed themselves without assistance, for their tongues have become abnormally shortened. For all these necessary affairs of daily life, they are completely dependent on their slaves. Once home from the slave-making raids, which they execute with such a dazzling disregard for danger and such military precision, they loaf about the nest, useless and idle. They polish their armour, perhaps, and sometimes beg their slaves for the food they cannot obtain for themselves. They are incomparable warriors, but they are parasites. Worse than that, they are parasites which lack the sense to preserve their means of life. Wheeler describes an experiment he made with

thirty of these Amazons which he had transferred
to an artificial nest. He provided them with a wet
sponge and a dish of honey, but in a few days the
helpless creatures were already suffering severely
from hunger and kept begging each other for food.
The only ants which did any better were two which
had chanced to fall head first into the honey dish.
Since their tongues were thus touching the honey,
they were able to lick it up.

These unfortunate Amazons soon began to die
of hunger in the midst of plenty and when only
sixteen survived, Wheeler decided to see if they
would adopt workers of another species. He therefore put three *subsericea* workers and three *nitidiventris* workers in with them. Both groups of
newcomers appeared frightened by their formidable
nest-mates, but the ants of the former species
seemed to be the more nervous of the two and
irritated the Amazons to a degree unfortunate for
themselves. One large Amazon in particular, who
had lost her right antenna and part of her right
front leg, seemed to be in a particularly vicious
mood, 'possibly,' writes Wheeler, 'because she had
been spending much of the day trying to comb an
imaginary antenna with an imaginary strigil and
had repeatedly tumbled over while attempting
this feat.' She made a vicious attack on two of the
subsericea workers and killed them. The third
managed to dodge out of the way, but by this time
all the Amazons had lashed themselves into such a
savage frame of mind that they even began to fight

among themselves. Meanwhile, the three workers of the other species, though apparently far from welcome, did not arouse the same aggressive fury in the Amazons and managed to survive unharmed apart from minor mishaps such as the loss of a leg.

Their good fortune did not last for long, however, for seven hours later two of them were found to be dead. There were now only two survivors, one of each species. The nervous *subsericea* worker kept out of the way and hid in corners but the other, obedient to her instincts, walked bravely up to one of the Amazons and putting out her tongue, she fed her. She then went to the honey dish, filled up her crop and returned to feed some of the others. She kept up her journeys, running to and fro between the honey and the Amazons, until all the ants, half-starved by this time, had been fed.

The next morning, the Amazons seemed to be sleeping together on the sponge with their adopted nurse in the midst of them, while the other worker tried to keep out of sight in a distant corner. Wheeler tapped the nest. At once the *nitidiventris* worker ran to the honey dish, drank in some honey and began once more to feed her charges 'like a solicitous mother who wakens early and sets about getting breakfast for a large family.'

Friendly relations seemed to have been established, and there were no incidents during the day to suggest that any of the former distrust remained.

Next morning the timid and outcast worker was found to be lying dead in a corner, but the other was as busy as ever feeding her formidable and, as one would have supposed, grateful dependants. The idyll was soon over. By noon, Wheeler observed that she was 'languidly dragging her body about the food chamber.' One of these intolerably stupid Amazons had bitten her through the head and she died a few hours later. They had killed their food supply.

However, Wheeler now put into the nest twenty large and active *subsericea* workers, together with a number of larvae and pupae. Whether it was the presence of their young or the confidence inspired by numbers is not clear, but these workers behaved very differently from their timid predecessors of the same species. The Amazons, apparently as incapable as ever of knowing when they were well off, rushed to the attack 'in a perfect frenzy of valour.' The workers held their ground, went over to the offensive and very soon the Amazons were in full flight with the new arrivals hard behind them. They seized the Amazons, 'showered them with formic acid, mauled them about, gnawed off their legs and left them in a pitiable plight,' a just retribution for their ungrateful murder of the devoted and kindly little ant who had worked so hard to feed them. Wheeler remarks that he watched their rout with much the same emotion as he used to feel when, as a boy, he read of the death of the suitors in the Odyssey.

A further stage in the demoralisation produced by slave-keeping is demonstrated in another species of ant-warriors, which still possess beautiful curved and pointed mandibles. Undoubtedly these would prove admirable weapons, if it were not for one insuperable difficulty. These ants have so little muscle-power that they are unable to use them. They manage to carry out slave-raids, however, by bluff. They seize the enemy in their formidable-looking mandibles, and, such is the power of suggestion, the threatened ants instantly drop the larvae or pupae they are carrying, although they are in no danger. These raids are bloodless of necessity and the parasite warriors are now only useful as show-pieces. The slaves are become the dominant ants of the nest.

4

GUESTS, PARASITES AND ENEMIES

Aɴᴛs, whatever the cause, appear to be passion-
ately devoted to the care of their young, though
they are not themselves mothers, and this amiable
feature of their character has placed them at the
mercy of a crowd of spongers and hangers-on who
exploit this instinct in a number of ways. Ants'
nests, in fact, frequently harbour a motley collection of 'guests,' some of which may be welcome
and desired, others a sore trial to their reluctant
hosts. They are generally divided into four groups—
the true guests, the indifferently tolerated lodgers,
the hostile, persecuted lodgers and the parasites.

The 'true guests' are generally beetles, and these
have become modified in their physical attributes
so as to get the best out of life among the ants.
Their tongues have become broad and short, so as
to receive liquid food more easily, their antennae
have become changed so that they can communicate
with ants and ask them for food, and they have even
grown 'handles' so that the ants can more conveniently carry them about.

They are great favourites with the ants, and some
of them, it has been maintained, appear to be kept

as pets. Other entomologists however, are in strong disagreement with this view.

Certain *Histerid* beetles—attractive dark-red creatures with tufts of golden-yellow hair—run about freely in ants' nests. This beetle is in the habit of sitting up and begging for food in much the same way as a dog, by raising the forepart of its body and holding up its front paws. When an ant passes near it, it waves its forelegs to attract attention. The ant stops and lays hold of the beetle with her front legs, raising it to an upright position. She then licks the beetle's face with a display of seeming affection. The beetle holds its head withdrawn inside its thorax for some time until at last the ant stops her licking and offers the beetle some regurgitated food with her tongue. At this the beetle puts out its head and quickly drinks it up. The ant then begins to lick its face again. The feeding and licking may go on alternately for a considerable time, 'as if,' writes Wheeler, 'the ant were fascinated with her pet and could not feed and fondle it enough.'

The beetle seems confident of its position among the ants and may even assume a 'ridiculous, cocky air,' as Wheeler says. Sometimes, a busy and preoccupied ant may knock it over without noticing it as it sits up and begs, but the beetle seems little put out. After a short struggle, it rights itself and sits up to beg again.

This beetle has the rather unpleasant habit of going to the refuse heaps and eating up the bodies of any dead ants which it may find there. However, it knows better than to offend its hosts. If any live ants come along while it is thus engaged, it submits with good grace to effusive fondling and licking and drinks in the regurgitated food it is offered without any sign of impatience. As soon as the ant is gone, however, it goes back to its unpleasing feast on its dead hosts' carcases.

Other kinds of beetle—particularly the *Lomechusa* beetle—seem to fascinate the ants to the point of folly and their presence may sooner or later mean the end of a colony. Such beetles are equipped with tufts of hair of a bright red or gold shade which the ants delight to lick. This is because they mark certain glands, which secrete an aromatic substance much appreciated by the ants. It is probably not a food so much as a luxury, but the ants find it so seductive that some of them become totally unable to resist it. They appear to become addicted to it, in fact. They lay aside their life of care and labour, and devote themselves to satisfying their appetite for this pleasure.

Such ants cannot be said to be actively condemned by their more conscientious sisters. There is no record kept of work performed or work neglected and the delinquent ants still have the characteristic nest smell which means that they are automatically accepted as members of the community. The community soon begins to suffer as a whole, however, by the presence, of these beetles, for their influence is subtly demoralising.

The substance they yield from their glands is so fascinating to the ants that they come to value the beetles which supply it to them to an altogether inordinate degree. When a beetle wants food, it taps an ant with its antennae or strokes its head with a forefoot. The ant finds herself quite unable to resist these blandishments and feeds the beetle. So much is harmless enough and might be done for any guest not altogether unwelcome. The ants also clean the beetles and seem to strive to entertain them by giving them rides on their backs. Again, such an indulgence is without serious consequences.

The ants' attention to these 'guests,' however, goes far beyond what might reasonably be expected of any host. For instance, the beetles lay their eggs at random round the nest, and the ants carefully gather them up and watch over them. The fat, white grubs which hatch out from them are fed and washed and cosseted to the neglect of the ants' own brood. Sometimes, in what must surely be regarded as an excess of hospitality, they even place these alien grubs on their own larvae and allow them to eat them up. Again, when danger threatens the nest, it is the beetle larvae which are picked up and rescued first in preference to their own brood.

Little by little, the broods of young ants begin to suffer from depletion and neglect. Fewer survive to reach maturity, and those that do are weak and may be pathological in form. These pathological

forms—'pseudogynes'—have some of the features
of a queen and some of a worker, but possess the
useful attributes of neither. It is thought that they
may derive from neglected and underfed queen
larvae, but this is not certain. However that may
be, they turn out to be feeble and lazy and altogether incapable of performing useful service to the
community. If they appear in large enough numbers, they may become such a dead weight on the
community that in the end prosperity declines and
the colony dies out altogether.

There is, however, one circumstance which may
ward off disaster and save the colony from becoming
fatally weakened. The ants appear to know that
their own larvae must be buried when they are
ready to spin a cocoon so that they may have walls of
earth close at hand to which they may attach their
threads of silk. As soon as they have finished their
spinning, however, the ants dig up the newly-
formed pupae and carefully clean off every particle
of earth. This office they also perform for the beetle
larvae with devoted diligence, but, although their
intentions are of the best, they thereby destroy them.
Like the ant larvae, the beetle larvae require to be
buried when they are ready to spin a cocoon and
transform themselves into pupae, but in this case it is
essential for their welfare that they should remain
covered with earth and not be dug up until the
fully-formed insect is ready to hatch out. The
solicitous ant, therefore, destroys the majority of
the beetle pupae by her well-meaning interference

and it is only those which she has forgotten, or failed to find, which come to maturity.

Other kinds of beetle are ignored or treated with indifference by the ants, but others again are anything but welcome visitors to the nest. These lurk about in dark corners ready to seize and murder any solitary ant as she hurries by. They may even creep up and kill the ants at night as they lie sleeping huddled together against the cold.

The ants, not unnaturally, dislike these assassins intensely and do their best to drive them from the nest. The beetles, however, are well equipped to defend themselves. When the ants come running up with open jaws, the beetle whips up its tail, thrusts its hind-portions into the ants' faces and emits a highly offensive vapour. The ants start back, apparently in considerable shock and discomfort, and the beetle then makes its escape without difficulty. The vapour seems to cause some injury to the ants who appear to be really ill after exposure to it. Some of them have been observed to be partially paralysed for some hours.

These beetles are classed among the 'hostile, persecuted lodgers,' although they seem to be more persecuting than persecuted. They must be a great trial to the ants, and probably cause more harm than most of the parasites which actually fasten themselves to their bodies. There are a number of these. The larva of one kind of fly, for instance, attaches itself to the neck of an ant larva and there forms itself into a kind of collar. Whenever its host is fed by worker ants, it uncoils itself and takes its share. If it becomes hungry in between times, it nips a neighbouring ant larva with its sharp jaws until it wriggles and so attracts the attention of the worker ants which come running up with food. It enters the cocoon with its host, but attaches itself to the cocoon wall and, when the empty cocoon is eventually thrown by the ants on to the rubbish heap, the fly emerges.

Certain mites attach themselves to the chins of full-grown ants, and these, too, secure food by reaching forward and sharing the meals of their hosts. This parasite goes even further, however. If it becomes hungry while there is no food about, it tickles its carrier with its forelegs and so induces the ant to regurgitate food for its own especial benefit. Donisthorpe relates that he has even seen this mite solicit another ant for food which it received at once.

This mite is not without consideration for its host, even though self-interest may play a part, for when the latter is feeding at the honey trough in an observation nest, it has been seen to move to one side so as not to be in the way. For some reason, it seems to prefer a young carrier and this para-site makes a practice of transferring itself to newly-hatched callow ants. The callows appear to be upset at first by its arrival and fall on their backs, roll on the ground and rub their chins desperately against anything they can find in a pathetic effort to free themselves. The mite shows considerable agility, however, in dodging about and keeping out of danger until at last the ant becomes resigned to her fate and sets about the normal routine of her life in spite of this unwanted companion.

Another kind of mite often makes its home among ants without arousing any resentment. It is a very agile creature and seems to use the ant as a form of transport, jumping on or off its back as though it were a conveniently passing bus. They do it very nimbly, somewhat after the manner of circus-performers, springing from the back of one ant to another while they are in motion. The ants, for their part, seem unconscious of the liberty which is being taken; or, at any rate, they show no resentment and generally give the impression that they do not even know that the mite is present in the nest. Occasionally, however, they may stroke the little creatures with their antennae and they have been seen to carry them to safety when the nest is in danger. They have reason to be tolerant, for the mites cause them no harm, eating neither ants nor their larvae, living or dead, but feeding instead on the nymphs of other mites.

Far more sinister from the point of view of the ant are those creatures which make their homes inside her body. There is a certain kind of *Phorid* fly which lays its eggs on the head of an ant. It tracks down an ant, probably by smell, since it has been observed to strike at a hand on which ants have been crawling, and when it has selected its victim, it pursues it with stealthy persistence. Gradually it draws nearer until it gets within striking distance. The ant, if she notices the fly in time, makes a rush to escape, and if she is too late, and the egg is implanted on her head, she may make the most frantic and pathetic efforts to free herself. They are seldom of any avail.

In time, larvae hatch from the eggs and eat their way into the ant's brain on which they feed. The doomed ant becomes a little lethargic, a little slow. The process goes on inexorably and the ant seems to become dazed and stupid. In the end her head is completely eaten away and the larvae, fat and full-fed, hatch out from the discarded remains.

The external enemies of ants are not as numerous as they might be, for ants have a notoriously unpleasant taste and few insects have any desire to eat them. Spiders for the most part leave them alone, although there are some species which feed on them. Some spiders, in fact, prey on ants by the exercise of histrionic skill, for they have acquired the trick of behaving exactly like an ant, even waving the front pair of their eight legs as though they were antennae and assuming a characteristically ant-like pose. The unwary ant runs up, expecting to find a sister, and is at once seized and eaten. The larvae of 'ant-lions' with their long necks and formidable jaws are amongst the most dangerous enemies of ants and trap them by digging pits in the ground. They lie at the bottom of a pit, entirely buried in the sand except for their menacing open jaws, and wait for an ant to slide down into them.

A certain Javanese bug with the impressive name *Ptilocerus ochraceus* exploits the ant's passion for aromatic secretions, as Jacobson has described. The bug stations itself at some place where ants frequently pass and as soon as an ant approaches, it raises up the front of its body so as to expose the site of the gland from which the secretion may be drawn. The ant runs up and at once begins to pull with its mandibles at the tufts of hair which mark the gland 'as if milking the bug,' and licks busily at the secretion which is freely exuded. Meanwhile, the bug folds its front pair of legs round the neck of the ant with a macabre appearance of affection. It is awaiting the right moment for the kill.

The ant keeps on licking at the gland, until the secretion gradually begins to produce a paralysing effect. She begins to draw in her legs and curl herself up. This is the sign for which the bug has been waiting and it acts at once. Its grip on the ant tightens, it thrusts its sharp beak into the ant's body and proceeds to suck it dry.

Although ants are not often eaten by other insects, they may be snapped up by a variety of reptiles and mammals—man himself eats ants in some parts of the world. And they are extremely popular with birds—a fact which, according to Gould, provides the main reason for their creation.

> The chief and most obvious Design of the noble Insect before us (he says), is its being intended as Sustenance for many species of Animals, but in particular for young Pheasants and Partridges. As soon as

the young Brood of partridges require Aliment, the fond Parent leads them to a Colony of Ants, where is a table spread for their Entertainment, and furnished with a sumptuous feast. . . The tender Infancy of these Birds calls for an early and delicious Repast, which is so happily contrived by the Disposition of Ants, as highly tends to exemplify the superior Wisdom and Beneficence of the great Creator.

It is to be doubted whether the ants appreciate the beneficence of this arrangement.

The greatest menace to the continued prosperity of ants lies in our own activities, for the environments which they need for survival are constantly being broken up and changed by the advances of civilisation. At times, too, we become actively their enemies, for when they make their nests in houses they turn into pests at once. Pharaoh's Ant, in particular, is a serious invader and extremely difficult to remove. Stories have been told of floors being pulled up and walls stripped in order to wipe out this persistent creature, and all without permanent effect. It is easy enough to massacre large numbers of the workers and to reduce them to a very low ebb, but the queen is always kept secluded in a place of safety and, so long as she survives, fresh workers are born to replace those that are killed.

On the other hand, so long as ants stay out of doors, they perform a useful function in aerating the soil by the construction of their nests. Since they are generally scavengers, they are also useful in clearing away the bodies of dead insects and attacking and devouring those which are disabled— among them of course, many insects which are harmful. On the debit side it must be admitted that they protect and preserve a few harmful insects for their own purposes, such as their 'ant-cows,' the greenfly.

5

THE LIFE OF THE MIND

WHEN writing or speaking of ants, there is a constant temptation to use words and phrases which would be more applicable to man. This is probably natural enough since, being social animals like ourselves, they undeniably resemble us in certain ways.

'In order to live in permanent commonwealths,' writes Wheeler, of the ants, 'an organism must be not only remarkably adaptive to changes in its external environment, but must also have an intense feeling of co-operation, forbearance, and affection towards the other members of its community.' This is a large claim to make for an insect, but the ant, so far as we can judge its character from its behaviour, appears to justify it.

Their co-operation one with another cannot be doubted for a moment. They divide up the labours of the nest among the workers to the common benefit, though how the tasks are allotted, how it is decided who shall do what, it is impossible even to guess. When a task appears too formidable for one ant to complete unaided, she does not hesitate to call on her sister ants for aid, and she receives it.

Certain tasks are undertaken by teams of ants working together and they appear to labour with complete understanding of what is necessary and what is to be expected from each.

Forbearance is another quality which ants appear to exhibit in their behaviour to an altogether unusual degree. Few, if any, naturalists have been rash enough to express any views about delinquent ants and it is by no means certain that they exist. Natural selection inevitably breeds ants for obedience and self-immolation on behalf of the community, for the independent-minded ant which wandered from the nest to lead a life of its own would be lost to the community and probably to life as well. At the same time, it is known that an ant may become addicted to the joys of licking at the aromatic secretions of beetles, and time spent in this manner is not spent on work. It is not known, however, that any sanctions are ever taken against such an ant—there is no police force in an ants' nest. The sick ant unable to work is also allowed to rest unmolested.

An extraordinary forbearance is certainly shown by ants to any alien creature that cares to make its home among them and is not actively hostile. Whatever the reason, all are accepted in time and most are fed and cared for as well. It has been observed that the ants may even lick and feed the mites which fasten themselves as parasites to their bodies, so unfailing appears to be their goodwill. Sometimes it happens that a colony of small ants makes its nest adjoining that of some large ants and they develop the habit of spending all their time in their neighbours' home. They lick their large friends, ride on their backs and constantly solicit them for food, which they receive, although they give nothing in return. Sometimes, however, the larger ants break into the nest of the smaller ants—breaking in is necessary since they cannot squeeze through the tiny passages used by their neighbours. At once, good relations are destroyed, for the smaller ants appear to be outraged by this intrusion and they defend their brood, which the larger ants want to take home with them, with passion and persistence. And such is the power which the sense of possession apparently gives them that they generally drive the large invaders away. Friendly relations are soon resumed, however, and the smaller ants continue to live on the charity of their larger neighbours, who still treat them with what seems to be faultless good humour and tolerance.

It has never been observed that ants of the same nest quarrel among themselves or invade each other's rights in any way. So close, in fact, is their mutual co-operation that they appear in many respects more like cells of one body than individual members of a community. An ant, isolated from its fellows, generally dies very rapidly.

Affection is a more difficult claim to establish on behalf of the ants. The mutual feeding which is so prominent a feature of ant life would suggest, at the least, a certain solidarity and it has been claimed that the attitude of an ant offering regurgitated food to a companion, with antennae thrown back, is somehow suggestive of extreme ecstasy. Maeterlinck has even suggested that it may replace the pleasures of sex which they are denied. Nobody can know. The world of an ant is so alien to our own that we can scarcely guess at the emotions they may feel. It is plausible, however, to suppose that they do experience emotions of a kind. Wheeler, possibly the greatest authority of all, says that they 'show unequivocal signs of possessing both feelings and impulses. In my opinion they experience both anger and fear, both affection and aversion, elation and depression in a simple, "blind" form, that is, without anything like the complex psychical accompaniment which these emotions arouse in us.'

Ants may claim mental qualities as well as emotions. There is no doubt, for instance, that they possess memory of a kind. When certain ants decide that the colony would be better off in another situation, and take active steps to remove their fellow nest-mates to the new site, they must presumably retain some mental concept of this other desirable locality to which they wish to move. The scouts of the Blood-red Slave-makers reconnoitre the district round the nest to be attacked days or even weeks before an actual raid takes place, and yet they not only report in some way the existence of a suitable colony for raiding, but retain such a clear picture of its situation that they are able to guide the army to it by a reasonably direct route. If all the pupae are not removed from the enemy nest on the day of the raid, they return on the next day to collect those which remain. They have retained in their consciousness overnight, one must suppose, the recollection of the pupae left behind.

Memory may not play a large part in helping ants to find their way from one place to another previously visited, for some ants appear to rely almost exclusively on scent and follow a trail blindly. Others appear to use intelligence or to possess some kind of steering mechanism, for they abandon the winding trail which they may have followed originally and go by the most direct route to their objective. Others, however, are helped by sight and, with these, memory must be assumed to have some importance.

Ants show foresight of a kind in that they not merely gather 'ant-cows' into their nests, but sometimes they bring in their eggs as well. As eggs they are no use to the ants at all, but the ants are aware—or at any rate act as though they were aware—that the eggs will one day develop into ant-cows and they keep them in their nests throughout the winter in this expectation.

In the use of tools, they appear to have made no progress, with one exception. The leaf-sewing ants might be said to use a tool to join one leaf to another and repair damage to their nests, even though the tool is—regrettably—their own larvae.

Ant society has solved many problems which confront man today, although the means by which they have solved them are hardly likely to seem attractive to us. Unlimited reproduction, for instance, is incompatible with stable conditions in the long run for other reasons as well as the obvious difficulties of food supply. Some kind of birth control appears to be necessary and an eminent American scientist has predicted that this will one day be achieved by diet. Such an idea would certainly be no novelty in the insect world and it is probably by diet restrictions more than anything else that the ants have succeeded in producing their toiling hordes of sexually undeveloped spinsters. The idea is singularly uninviting and it cannot be supposed that the American scientist had such an arrangement in mind when he made his prediction. But it *is* a solution and it has certainly worked among the ants.

With the exception of the queens and the short-lived males, ant society has established itself on strictly egalitarian lines. There are different castes, it is true, and sometimes they are rigidly differentiated one from another. Yet all must work, though the tasks be different, and none may claim a greater degree of comfort, a larger share of the fruits of labour, than any other. A. D. Imms has said that the community is 'virtually a vast proletariat of sterile working classes.' Although there is no compulsion, so far as is known, each member of these classes toils all day and sometimes finds no rest from labour even at night. And they do this for the bare reward of food and shelter, at the sacrifice of individuality and all individual ambitions. The state, the colony, is all-important, and each member can claim only a borrowed importance as a part of that whole. And when the whole is threatened, the parts, the individual members, must sacrifice their lives in its defence, as they have already sacrificed their energy, their talents and all their strength.

It is a sombre and a pathetic picture, and for us it is not without sinister implications. None the less, in watching the ceaseless scurryings of the ants in their nests, it is impossible not to feel some measure of sympathy, or even affection, for these little creatures. When we contemplate a cold and solitary beast like the spider, ensnaring her prey by careful cunning and consuming it with un-hurried cruelty; a creature which knows so little of the natural loyalties that she may kill and devour her mate before the act of mating is complete; a monster which feels so little regard for her off-spring that she may slay and devour them, too, if they fail to leave her at the exact moment when the processes of her physiology demand it; when we contemplate this strange being, we feel at a loss. We cannot hope to understand her.

But the ant is different. She may sometimes be aggressive and cruel, it is true, but she can generally plead provocation. And it is not her nature. In most of her ways, she seems to show herself gentle and benevolent almost to a fault. Her loyalties are directed, as ours are, towards her own family and her own kind. Even her amusements and her indulgences are not completely alien to our own. Above all, she wins her bread and makes her way in the world by activities which are so recognisably similar to many of our own, that we can scarcely withhold a sympathetic interest.

The ant is vastly more experienced in the complexities and problems of social life than we are. Over a hundred million years ago, when our remotest ancestors had scarcely crept from the slime, the ant was not so very different from what she is today. In our quick surge to supremacy, we have already advanced far beyond her in the arts and uses of civilisation, and it may be that our progress will bring her empire to an end. We shall do well, however, if our own achieves the stability which hers has shown, and endures throughout the passing of so many aeons of time.